Appraising Machinery and Equipment

"All values are of the nature of forecasts of events and are subject to the uncertainties of all prophecies. Values fluctuate with changes in prevailing opinions of what the future is likely to bring. They can never be determined by formulas or computations alone."

From Anson Marston, Robley Winfrey, and Jean C. Hempstead, *Engineering Valuation and Depreciation,* Iowa State University Press, Ames, Iowa, 1968, p. 7. Reprinted by permission.

Appraising Machinery and Equipment

Machinery and Equipment Textbook Committee
of the American Society of Appraisers

John Alico, P.E., C.C.E., F.A.S.A.
Editor

McGraw-Hill Publishing Company
New York   St. Louis   San Francisco   Auckland   Bogotá
Caracas   Colorado Springs   Hamburg   Lisbon
London   Madrid   Mexico   Milan   Montreal
New Delhi   Oklahoma City   Panama   Paris
San Juan   São Paulo   Singapore
Sydney   Tokyo   Toronto

Library of Congress Cataloging-in-Publication Data

Appraising machinery and equipment/the Machinery and Equipment
  Textbook Committee, the American Society of Appraisers; John Alico,
  editor.
     p. cm.
  Includes index.
  ISBN 0-07-001475-2
  1. Industrial equipment—Valuation—United States. I. Alico,
John. II. American Society of Appraisers. Machinery and Equipment
Textbook Committee.
HD39.35.A66 1989   88-13226
658.1'5242—dc19   CIP

Copyright © 1989 by the American Society of Appraisers. All rights
reserved. Printed in the United States of America. Except as permitted
under the United States Copyright Act of 1976, no part of this
publication may be reproduced or distributed in any form or by any
means, or stored in a data base or retrieval system, without the prior
written permission of the publisher.

234567890 DOC/DOC 895432109

ISBN 0-07-001475-2

The editors for this book were Martha Jewett and Barbara B. Toniolo, the designer was Naomi Auerbach, and the production supervisor was Suzanne Babeuf. This book was set in Baskerville. It was composed by the McGraw-Hill Book Company Professional and Reference Division composition unit.

Printed and bound by R. R. Donnelley and Sons Company

For more information about other McGraw-Hill materials, call 1-800-2-MCGRAW in the United States. In other countries, call your nearest McGraw-Hill office.

Dedicated, with gratitude, to those pioneers who contributed to the development and progress of machinery and equipment appraising. Their teachings and contributions to the literature of this discipline helped to establish the importance of evaluating machinery and equipment as personal, tangible property and a basic element of the national economy.

Contents

Contributors   ix
Preface   xi

1. **Valuation Theory and the Machinery and Equipment Appraiser**   1
   Kenneth A. Martin, FASA, *Arthur Andersen & Company, New York, New York*

2. **Classification of Property**   9
   John J. Connolly, III, ASA, *Nationwide Consulting Company, Inc., Fairlawn, New Jersey*

3. **Identification of Machinery and Equipment**   17
   Alan C. Iannacito, ASA, *ACI Associates, Denver, Colorado*

4. **Purposes of Appraisals**   29
   David M. Graham, ASA, *Greenbank, Washington*

5. **Replacement Cost New Concepts**   39
   Merritt Agabian, ASA, *A&M Appraisal Company, East Walpole, Massachusetts*

6. **Sources of Pricing and Reference Material** 49
Paul Rice, ASA, *Arthur Andersen & Company, Los Angeles, California*

7. **Depreciation Theory** 59
John Alico, P.E., FASA, *Alico Engineers and Appraisers, Birmingham, Michigan*

8. **Fair Market Value Concepts** 79
Robert S. Svoboda, ASA, *American Appraisal Associates, Inc., Milwaukee, Wisconsin*

9. **Liquidation Value Concepts** 129
Leslie H. Miles, Jr., ASA, *MB Valuation Services, Inc., Dallas, Texas*

10. **Insurable Value** 153
Kal Barrow, ASA, *Arthur Andersen & Company, New York, New York*

11. **Scrap/Salvage** 161
David M. Graham, ASA, *Greenbank, Washington*

12. **Value-in-Use versus Value-in-Exchange** 167
Leslie H. Miles, Jr., ASA, *MB Valuation Services, Inc., Dallas, Texas*

13. **Appraisal Report Content** 175
George D. Sinclair, FASA, *Keystone Appraisal Company, Philadelphia, Pennsylvania*

14. **Ethics** 191
John Alico, P.E., FASA, *Alico Engineers and Appraisers, Birmingham, Michigan*

Index   201

Contributors

Merritt Agabian, ASA — President, A&M Appraisal Company

John Alico, P.E., FASA — President—Alico Engineers & Appraisers

Kal Barrow, ASA — Manager, Appraisal and Valuation Services, Arthur Andersen & Company

John J. Connolly, III, ASA — Secretary/Treasurer, Nationwide Consulting Company

David M. Graham, ASA — Greenbank, Washington

Alan C. Iannacito, ASA — President, ACI Associates

Kenneth A. Martin, FASA — Manager, Appraisal and Valuation Services, Arthur Andersen & Company

Leslie H. Miles, Jr., ASA — CEO, MB Valuation Services, Inc.

Paul Rice, ASA — Manager, Appraisal and Valuation Services, Arthur Andersen & Company

George D. Sinclair, FASA — President, Keystone Appraisal Company

Robert S. Svoboda, ASA — Senior Engagement Manager, American Appraisal Associates, Inc.

Preface

The American Society of Appraisers published, in 1969, the first work entirely devoted to the principles and procedures of appraising machinery and equipment. The subject was presented in the second of a series of monographs published by the Society to contribute to the knowledge of appraisers, and to strengthen the status of the profession. Prior to that time, the complexity of the "M and E" concept and practice was emphasized by the fact that so little generic material had been published.

In the interim, ASA has presented a series of courses and seminars for students, appraisers, lawyers, accountants, assessors and other related professionals at various colleges and universities throughout the United States. These educational efforts were supplemented by lectures and presentations given during the annual meetings of the American Society of Appraisers held in a number of cities throughout the United States, Mexico, and Canada.

In 1983, the need for organizing a Machinery and Equipment Committee to establish and structure a long-range program for implementing the Society's educational goals became apparent. Under the leadership of Alan C. Iannacito, a textbook subcommittee was formed and eleven members, all specialists and practitioners in the field of "M and E" appraising, accepted chapter assignments. The "consortium" approach was, in the opinion of the Committee, the best practical solution to presenting the most comprehensive coverage of the subject.

In the Introduction to *The Appraisal of Machinery and Equipment*, ASA Monograph No. 2, published by the American Society of Appraisers, 1969, Dexter D. MacBride, FASA, J.D., then chairman, International

Publications Committee, and later, executive vice president of the Society wrote, prophetically:

> It is recognized in the "consortium" approach employed by this monograph that certain concepts are repeated, procedural steps are re-emphasized, and certain adjurations are restated. Such an "overlapping" is considered, in the educative process, to be a valid means of communication. If repetition provides a useful mnemonic instrument, the Society's goal of effective instruction will be in part achieved.
>
> Of greater concern, perhaps, will be "differences" which may be discovered in Approach and Procedure, in consequence of the changing, interacting patterns of local and regional custom, legal jurisdiction, constantly shifting economic emphasis and emerging professional criteria, as seen through the eyes of the several authors.
>
> Illustrative of the problem of "differences": in the seemingly definite environment of Law, not only do express differences exist because of jurisdiction, but there is the fact that what "the Law" requires to be done Today, it may require to be done differently, or not done, Tomorrow.
>
> Law has only the certitude of Yesterday; to be a vital living instrumentality of guidance and service, Yesterday's rule must be challenged by events of Today, and changed to meet problems of Tomorrow.
>
> It is not surprising that the Appraisal profession, like "the Law," must be amenable to the flux of Change. The science of "placing a value" in terms of a unit of measurement (as the dollar is generally assumed to be) is plagued with notions of exactitude. A moment's reflection brings the problem into focus: the dollar unit, although assumed by many to be an exact "yardstick" of measurement, is itself subject to fluctuations, and trembles in the winds of national crises (rumors of war, monetary re-evaluations in other governments of the world, etc). Our six base categories of measurement—temperature, mass, length, luminous intensity, time, electric current—are represented by precise units. Money is not represented in so exact a fashion.
>
> The skilled appraisal practitioner, knowing such "boundaries" as Law and Money Measurement are constantly shifting, will not be surprised that the varying experience-exposure of experts will reflect a multiplicity of concepts, a variety of procedures. These will be understood to be congruous, but not identical. They may reflect areas of difference, as does the environment which provides the data. With an appreciation for these factors, it is our hope this monograph will be accepted by the reader for possible use as a guidepost, testing ground or catalyst.
>
> It should be noted that those responsible for writing and editing this monograph are well aware the publication represents a "beginning" only. The "Appraisal of Machinery and Equipment" is but an introduction, a "view and inspect" of a house with many rooms. There is much to investigate, analyze, explain, report. With this initial presentation, the Society welcomes additional commentary from skilled practitioners who will undertake to add to the knowledge of the profession.

If the reader will substitute the word *book* for *monograph* he or she will find that this passage is still timely and can directly relate to this new book.

MACHINERY AND EQUIPMENT COMMITTEE
CHAIRMAN 1986–1988: John Madge, ASA
MEMBERS: Merritt Agabian, ASA
Ralph C. Alberti, ASA
John Alico, FASA
Kal Barrow, ASA
John J. Connolly, III, ASA
Alexander J. Cornett, Jr., ASA
Alan C. Iannacito, ASA
William F. Jacobs, ASA
Kenneth A. Martin, FASA
Leslie H. Miles, Jr., ASA
H. Denis Neumann, ASA
Paul Rice, ASA
George D. Sinclair, FASA
Robert S. Svoboda, ASA
William E. Widener, ASA
Edward A. McInnes, FSVA (International)

1

Valuation Theory and the Machinery and Equipment Appraiser

Kenneth A. Martin, FASA

Manager, Appraisal and Valuation Services, Arthur Andersen & Company, New York, New York

This text, as a whole, covers the technical theory, principles, and methods which must be understood to appraise machinery and equipment. However, it is important to explore briefly the broad theoretical question, "What is value?"

What Is This Thing Called Value?

Appraisers focus on one type of value, i.e., economic value. Actually, the concept of value comprises many influences. These influences are physical, psychological, philosophical, legal, political, and economic. All these play an important part in the concepts of ownership and property and how a person conceives value.

Fundamental Types of Value

These influences have been classified into 10 categories of human values by Edgar Sheffield Brightman, in *Preface to Philosophy*.[1] Brightman cautioned that "The variety of experiences of value is so great that no one classification of them would be final." However, to develop an insight into the complexity of the question, "What is value?" he proposed the following list.

I. *Purely instrumental values*
  A. *Natural values.* [These include] the forces of nature—life, gravity, light—insofar as they operate causally and are accessible to all. The intrinsic values to which natural values give rise apart from control by purpose are usually bodily or aesthetic.
  B. *Economic values.* [These include] physical things, processes (like power), human labor, or services, insofar as their possession is a socially recognized property right, acquired or surrendered by exchange for equivalents or supposed equivalents. Economic value is exchange value. One who regards economic values as intrinsic is a miser. But abundance or deficiency of economic wealth has a profound effect on both the quantity and quality of realizable intrinsic values. Money, the symbol of exchange, has been called *coined life*.

II. *The lower intrinsic values.* (This group is called "lower" because its values are narrower, more partial, than the "higher" ones; they include a smaller area of value experience and are more dependent on other values for their own worth.)
  A. *Bodily values.* These are not to be confused with the natural instrumental values, which are purely causal. Bodily values include only the enjoyment in consciousness of the well-being resulting from satisfactory bodily functioning. The feeling of being in good health, the joy of living, the pleasures of sex, and the delight of successful athletic endeavor all belong in this group. Bodily values constitute only one limited realm of intrinsic value experience, but they are instrumental to an incalculable amount of weal and woe among the higher intrinsic values.
  B. *Recreational values.* [These include] the satisfactions that come from play, humor, or mere amusement. These are the chief values of childhood, but are essential to the healthy mind at every age. Since their instrumental value is great, some regard them as exclusively instrumental; but the very instrumental value of recreation is lost on one who does not enjoy it as intrinsic. Yet recreation is not the serious business of life and offers a relatively narrow range of experience.
  C. *Work values.* Just as play is joyful, so work should be. The mere fact of being employed is itself a satisfaction. The production of instrumental values is itself an intrinsic value, or would be in a reasonably just economic order. Yet satisfaction in usefulness is a very slender value by

---

[1] R. E. Hoopel et al. (eds.), *Preface to Philosophy*, Macmillan, New York, 1947, pp. 136–139.

itself. Its justification is chiefly beyond itself in the intrinsic worth of what is being produced.

III. *The higher intrinsic values.* (This group is called "higher" because its values are broader, more inclusive of experience as a whole, more independent, and more coherent. It is impossible to group these higher values in a scale of increasing excellence, except that social values are intrinsically lower than the others.)

A. *Social values.* This term refers to the special value that is experienced through the consciousness of association, cooperation, or sharing. It is clear that many of life's most highly prized values can be experienced only this way. Social values should be called "higher" if only because they embody the worth of personality. Every true value is enhanced when experienced as a social value.

B. *Character values.* This somewhat unsatisfactory term designates the experience of a good will, the conscious choice of what is believed right and best. The word moral is avoided because the moral life is not merely the good will but also actual organization of the whole experience of value by the will. Thus morality is the experience of the whole table of values, while character values refer exclusively to the act of choosing. A good character is indeed a jewel that shines by its own light and is respected by every rational mind, but it is not the only higher value. Yet it is so necessary that without the control by good will all the other values soon become disorganized, incoherent, and self-destructive.

C. *Aesthetic values.* The values of aesthetic satisfaction include not only the beautiful but also the sublime, the tragic, the comic, and many other gradations. The aesthetic, whether in nature or in art, is an experience in which there is, or at least appears to be, an adequate expression of purpose in such a way as to stir feeling and achieve harmony. Art may be defined as the conformity of expression to purpose. Character value is independent of success in achieving what is chosen; aesthetic value depends entirely on such success. What does not embody the intended meaning is not aesthetically adequate. That aesthetic values are intrinsically satisfactory is the universal testimony of humankind. Like character values, aesthetic values are experiences in which the whole of life is mirrored or organized from a special point of view.

D. *Intellectual values.* The intellectual values are the experiences of truth loving and truth finding. It remains an empirical fact that much truth is valued by the noblest spirits. No one could deny that the truths of science and philosophy are instrumental to the control of inner and outer experiences; it is a familiar observation that intrinsic values are also instrumental. The joy of knowing, the mere satisfaction of curiosity, the "wonder," which according to Plato and Aristotle is the beginning of philosophy, are experiences common to every human being. These constitute the intrinsic aspects of the intellectual values.

E. *Religious values.* [These are] the values which are experienced when people take an attitude toward the value experience as a whole and toward its dependence on powers beyond themselves. Insight into this dependence elicits feelings of reverence and acts of worship. The special quality of the whole which is deemed worthy of worship is called holiness. Like all other higher values, religious values are an organization of the total value experience from a special point of view. Social values organize the whole from the standpoint of sharing; character values, from the standpoint of control by will; aesthetic values, from the standpoint of appreciative feeling; intellectual values, from the standpoint of knowledge; and religious values, from the standpoint of worship of and cooperation with the objective cosmic source of values.

As humankind experiences all these values, a series of attitudes develops. In the formation of social order, these attitudes are transformed into systems of religion, sociology, economics, law, and politics. These symptoms of social order are developed to protect personal physical survival and social esteem and to enjoy personal physical pleasures and/or power. It then follows that humankind perceived that there was something of value to protect or enjoy. This perception of having something of value is the basis of ownership.

The Concept of Ownership and Its Relation to Value

In his paper, "On the Prehistorical Origins of Ownership," Professor Jean Canonne says,

> Central to the question of the origins of attitudes toward ownership is the interdependence between value and the status which property conveys to its owner. These attitudes appeared once our distant ancestors, through a system of time-factoring, passed from the prehensile stage to the stage of occupation, whereupon the physical and spiritual satisfaction they derive from meeting their vital needs was enhanced by means of social concept that was deliberate, communicable and understandable by the group.[2]

The development of attitudes has its basis in past experience or knowledge of past experience. People learned that the things they value not only have value to the owner but also may be desired by others. History books are full of the constant struggle between those who own something and those who desire it. Canonne says, "It is obvious that value cannot survive without property which, in turn, receives its entire justification from the pre-existence of value.[3] Ownership therefore is a situation whereby someone possesses something that he or she considers valuable.[4]

---

[2]Jean Canonne, "On the Prehistorical Origins of Ownership," *International Real Estate Journal* (The International Institute of Values), 1985, vol. 7, no. 1, p. 67.

[3]Ibid., p. 66.

[4]It should be noted that this statement contains a paradox in that value is a state of attitude that is subjective, whereas property, the something owned, is identifiable and objective; yet each is dependent on the other.

The desire for ownership led people to develop things useful to themselves even though the things have no utility in their natural state. Once the things are no longer useful to the owners or users and the desire to possess them no longer exists, the items are disposed of. The act of using and/or disposing of something is a situation where only the one who has possession of something can exercise this right. Recognition of this fact has developed into the concept of a "bundle of rights" that attaches itself to the property. This bundle of rights allows the owner to use, exploit, keep, or dispose of the thing owned. In exercising these rights, the owner must decide if ownership has any short- or long-term advantages.

Since the relationship of value to the thing owned exists in the mind of the owners, it is the owners' measurement of the anticipated sum of future value services that the thing will procure that forms the owners' decision to exercise their rights. Therefore, value in its basic terms is simply an opinion.

This concept raises another paradox in the relationship between ownership and value, since value is derived only from the anticipated future utility of the thing owned, while the thing owned has its origins in the past. (It must exist or there would be no rights to exercise.) Therefore, value and ownership are linked to a knowledge of the past that is projected into the future but is measured in the present.

The Concept of Property

The discussion of the concepts of the relationship of ownership and value leads to questions on property. What is property? What types of property are there? How is property classified for appraisal purposes? What are the value characteristics of property?

Property is something that is owned. This something can be owned by one person, co-owned with one or more persons, or owned by corporations, by institutions or, in the public interest, by government. This something owned does not necessarily have to be tangible since value is subjective on the part of the owner.

Since ownership, regardless of its form, is the exclusive right to possess, use, and/or dispose of the something owned, the appraiser does not really appraise a property. The appraiser appraises the rights of ownership.[5] As an example, machinery and equipment installed in a factory and subject to a loan have two owners. Both the lender and borrower have interests in the property. Therefore, the machinery and

---

[5]These rights are commonly referred to as the interest a person has in a property.

equipment are not property; the rights (interest) associated in the ownership constitute the property. In this case, we see that rights are transferable. Each owner has rights to the same property. Each is using the same property in a different way: the lender to earn interest and the borrower to manufacture something. Each has made a value decision that allows each of them to exercise rights by using a thing to satisfy their desire.

Therefore, the machinery and equipment are not the property. The rights to use constitute the property.

Property is not necessarily made up of one component. The rights in the preceeding discussion refer to the use of many items: machinery and equipment. When a going concern is valued, we see that many components are found, i.e., land, improvements, buildings, machinery, and goodwill. Each component has its own bundle of rights, but when all components are part of a business, the business has its own bundle of rights, which may or may not be the sum of the components.

In *Appraisal Principles and Procedures,* Henry Babcock points out that

> Each of the components in an integrated whole property has in general, two different values, depending on whether it is considered as integrated with (not separated from) the whole or whether it is considered as separated from (or independently of) the whole. If the component is considered as separated from the whole it is treated by itself, ignoring its relationship to other components, and it is called a fraction. These technical definitions are of the utmost importance in the exposition and application of valuation principles.[6]

Classification of Property for Valuation Purposes

The rights to property can be classified for valuation purposes as the benefits derivable from two broad classes of property: investment property and noninvestment property.

Investment property is defined by Babcock as "one which is produced, acquired, or held for the sake of monetary income or monetary profit." Noninvestment properties are "those which do not possess this characteristic of generating monetary returns, but which are of such a nature that the benefits of ownership are derived by use and/or consumption by the owner."[7]

---

[6]Henry Babcock, *Appraisal Principles and Procedures,* 2d ed., American Society of Appraisers, Washington, D.C., 1980, p. 54.

[7]Ibid., p. 74.

Within these two broad classifications, the property can be either marketable or nonmarketable. Marketability is dependent on desirability, which in turn is dependent on how others project the future benefits derived from the utility of the property.

It should be noted that a third classification is personal property. Personal property has only owner value. The value involves only one person, who has an internal knowledge of the future possibilities for its use although it may or may not have an economic value.

Since personal value involves only one person, appraisers cannot measure this value. Appraisers measure impersonal value. Impersonal value is defined by Canonne as "A communicable representation, which may or may not be a modification of personal value; it involves more than one person, is social and characteristic of achieved man."[8]

Investment property has all three value characteristics. It has investment value since it can be produced, acquired, or held for the sake of monetary income or monetary profit. It has market value since the rights are transferable through the sale of the property, and it has owner value since the owner can also derive personal pleasure through the use and/or consumption of the property.

Noninvestment property, if it is marketable, has two value characteristics: market value and owner value. If it is not marketable, it has only owner value.

Value Characteristics of Machinery and Equipment

Machinery and equipment can have investment value since they can be produced, acquired, or held for the sake of monetary income or profit. The investment value is usually found when they are considered as components of an integrated whole property. This is identifiable in the valuation of a going concern and when allocating purchase/sales prices. Investment value would also exist in the eyes of machinery and equipment manufacturers, dealers, and lessors.

In most cases, machinery and equipment are appraised fractionally (independently of the whole). When appraised in this manner, they are a marketable noninvestment property. Their value is not dependent on their future sale; therefore they are not held as investment property.The value characteristics are their market and owner value.

Machinery and equipment would be classified as nonmarketable noninvestment property when their only value attribute is owner value.

---

[8]Canonne, "Prehistorical Origins of Ownership," p. 70.

This nonmarketable property constitutes items specially designed and built to order which by themselves have no value except for their present use.

Summary

The concept of value is made up of a rather complex set of experiences that have been classified into a table of human values. Common throughout this table of human values is the perception that the owner has something of value. This then forms the basis of ownership. It was pointed out that value is an opinion and that value explains property, which is itself the basis of value. Value does not exist where there is no property or its representation. The opinion of value is formed on past experiences and knowledge of the utility of the thing valued, but it has no basis until a knowledge of the future possibilities is established. Hence, the definition of value is the anticipated sum of future services that a thing will procure. In present appraisal terms this definition is stated as "the present value of future benefits."

In exploring the concept of property, we find that, simply defined, property is something owned. It can be either tangible or intangible and can be solely owned or co-owned. We also find that what is owned is actually the bundle of rights to possess, use, and/or dispose of something owned. These rights represent the owner's interest. For valuation purposes, property is classified in two broad categories—investment property and noninvestment property—that have value characteristics. These value characteristics, as they apply to machinery and equipment, are investment value, market value, and owner value.

The appraiser's knowledge must be broad. The appraiser is interpreting value that has its roots in a complex system of physical, psychological, philosophical, political, and economic experiences. The appraiser works within a framework of social order that is in constant change. Laws change, the economy expands and contracts, and there are obstacles of social and technological change. All this must be evaluated in terms of the past, present, and future.

The machinery and equipment appraiser is not a scientist but a professional who uses scientific methods to measure the utility of property and its ability to produce goods, services, and benefits to the owner or the potential owner through the use of present-day technical evaluation methods.

The following chapters present both the theory and application of these technical evaluation methods as they apply to the appraisal of machinery and equipment.

2
Classification of Property

John J. Connolly, III, ASA
Secretary/Treasurer, Nationwide Consulting Co., Inc., Fairlawn, New Jersey

For purposes of this text and generally speaking, property can be divided into three distinct categories: real, personal, and intangible. There can be subcategories of the major divisions, including the classifications of real property such as raw or developed land, residential or multiple family housing, or commercial and industrial buildings.

Definition of Real Estate

Real estate is "the physical land and appurtenances including structures thereto."[1] Real estate is "immobile and tangible"; that is, it has substance, usually immediately perceived. Exceptions to immediately identifiable things include the rights attributable to the real estate itself, including mineral rights, leaseholds, and easements.

The distinction between *real estate* and *real property* is that real property includes the interests, benefits, and rights inherent in the owner-

---

[1]Reprinted by permission from *The Appraisal of Real Estate*, 8th ed., p. 8. Copyright © 1983 by the American Institute of Real Estate Appraisers, Chicago.

ship of physical real estate. Some state statutes hold that real estate and real property are synonymous.[2]

Consider, however, that in some states an obvious piece of equipment, such as a mine hoist affixed to the property via the foundation, may be considered by the state to be real property. On the other hand, a grinding mill attached by a foundation to the same mine property may be (and probably is) considered machinery and equipment personal property.

If there is a question in the appraiser's mind regarding the classification of real property or personal property, state law should be researched. The appraiser should know the state laws and note the reference(s) to specific citations of the statutes as they pertain to kinds of property.

Definition of Personal Property

Personal property is defined as those tangible items that are not permanently affixed to real estate and can be moved. A general definition of personal property is anything and everything, excepting intangibles, that is not realty and/or not permanently attached to the realty.

Examples of Personal Property

Equipment, for example, could include office furniture and machines, storage shelves, suspended fluorescent lights, manual and powered bench tools, floor-type water coolers, drapes, electric range-sink-refrigerator combinations, wall-mounted cabinets, hoists and monorails, stock storage mezzanines, exterior security lights, and modular office dividers.

It does not include the land, land improvements, or building and heating elements which are attached to and made part of the structure. Specific items considered as real estate by the appraiser could include acoustical ceilings, recessed fluorescent lights, wall paneling, wall-to-wall carpeting, chain link fencing, metal awnings, and stock racks on exterior walls.

Further distinction of property classes can be made in difficult cases according to the manner in which the property is affixed (if attached at all), the intention of the parties affixing the once personal property to

[2]Ibid., p. 8.

the real estate, and the purposes for which the premises are used. "Generally, items remain personal property if they can be removed without serious injury either to the real estate or the item itself."[3]

Fixtures

A fixture is an asset that lies in the gray area of classification. A *fixture* can be defined as an asset that normally would, by itself, be considered personal property. However, because of the manner of its installation, the item has become affixed to the realty, or it may be utilized for an extended period of time, in place, and has thus become part of the real property. If the fixture is removed, impairment may result to the real property.

Court opinions are occasionally required to make the distinction regarding certain properties where such situations exist. This has resulted in the formulation of the "law of fixtures." The law of fixtures defines a fixture as "an article that was once personal property but that has been installed in a structure or attached to land or a building in some more or less permanent manner, so that such an article is regarded, in law, as part of the real estate."[4] Items such as ceiling lighting and certain built-in cabinetry are considered real property.

Although fixtures are classified as real estate, trade fixtures are not. A *trade fixture*, also called a *chattel fixture*, is an article owned and attached by a tenant to a rented space or built for use in conducting a business. Therefore, trade fixtures are not considered to be real estate and are not endowed with the rights of real property.

Special Cases

There are instances when items may logically fit under the real estate improvement category but are appraised as machinery and equipment. An example could be an air conditioning system. If the appraisal project is being conducted for a potential allocation of purchase price, the air conditioning system needs a separate value because it depreciates faster than the building. To explain this rationale, we can look at this same asset in reverse. A building is purchased and set up for a 30-year

---

[3]Reprinted by permission from *Real Estate Appraisal Terminology* (rev. ed.), p. 184. Copyright © 1984 by the American Institute of Real Estate Appraisers, Chicago.

[4]Ibid., p. 184.

depreciation. Approximately one year later, window air-conditioning units are replaced by a central system. This new air conditioner would be capitalized and set up for a 7- or, perhaps, even 10-year life, which is realistic. The appraiser must give this same realistic benefit to a new purchaser for an allocation rather than include it with a building to carry over the possible 30-year life that would not be realistic to the air conditioner. In this case, the purpose of the appraisal may govern the inclusion or exclusion of items in keeping with state or federal laws.

Does this mean that a machinery and equipment appraiser cannot accurately value an item because it has been classified as real property? Definitely not. In most instances, a real estate appraiser has nominal, if any, experience in the evaluation of personal property and, therefore, may not be able to reflect accurately the item's value. The same could also be true on the reverse side. When an item of real property is considered personal property, the machinery and equipment appraiser should not attempt to value it without the expertise to do so. An example of when this may occur in Pennsylvania is in the classification for property tax purposes of certain process buildings which are not considered real estate or real property but rather personal property. Most machinery and equipment appraisers do not have the expertise to determine the market value of these buildings and should, therefore, seek outside help to solve this problem.

Intangible Property

By definition, *intangible assets* are "such values which accrue to a going business such as goodwill, trademarks, patents, copyrights, franchises, or the like...a noncurrent asset which exists only in connection with something else, as the goodwill of a business."[5] In this definition, the emphasis is placed on goodwill and little is implied for equally important intangibles, such as a skilled labor force, engineering drawings, contracts, or the quality of management. Goodwill is not an amortizable asset in the rules and regulations of the Internal Revenue Service. Business enterprise appraisers are more interested in the intangible assets that have a supportable life. This is so that this kind of intangible property may be amortized for tax purposes.

It is of the utmost importance that the personal property appraiser work closely with the real estate appraiser if a total property is being

---

[5]Joseph R. Nolan and Michael J. Connolly (eds.), *Black's Law Dictionary*, 5th ed., West Publishing Company, St. Paul, Minnesota, 1979.

appraised. An agreed upon *division of coverage* would prevent any duplication of value in overlapping areas where both the real and personal property appraiser could pick up the same items and include them in their evaluation. Some gray areas where this overlapping may occur in industrial plants could be in the following examples: craneways, foundations, electric power, and boilers (especially process boilers).

Effect of State Laws on Personal Property Classification

When an appraisal is being made for property tax purposes, the personal property appraiser must be cognizant of the applicable laws and classifications within that state. A good example of this is in New Hampshire where, as a property appraiser, you are assigned to appraise a ski area. In New Hampshire, the entire ski lift operation is taxable as real estate because it is intimately interwound with the primary use of the land, i.e., skiing. Therefore, all the peripheral equipment—cables, towers, chairs, T-bar lifts, and grips—would be considered real property by the state and, therefore, taxable as such. For depreciation purposes (federal law) there may be the same situation as in the air-conditioning example described earlier in this chapter.

In New Jersey, any machinery and equipment or other personal property put in use after 1978 is exempt from taxation; that is, machinery and equipment are not considered as fixtures.

Indiana has a real and personal property guide. The guide states that the use of a questionable unit of machinery, equipment, or structure will determine its classification as real or personal property. If the unit is used directly for manufacture or a process of manufacture, it is considered personal property. If the unit is a land or building improvement, it is considered real estate. Some examples of items that are considered personal property in Indiana are as follows: aluminum pot lines, bowling alleys, control booths, cranes and crane runways, dock levelers, irrigation equipment, yard lighting, mixers and mixing houses, silos, steam/electric generating plants and equipment, stone-crushing plant equipment, all substation equipment, tanks used as part of a manufacturing process, and power wiring. All these items are considered personal property.

An experienced machinery and equipment appraiser may have an assignment that will include a variety of types of personal property. Quite often the assignment will ask for an inventory and valuation of production equipment, material handling equipment, maintenance equipment, office equipment, shop equipment, and office furniture and fixtures. The appraiser is correct in appraising all the machinery and equipment in the facility, rather than selecting a few or trying to determine their classification for this assignment. If later the appraiser is informed that some items do not belong in the personal property category, it is a simple matter to take out those items. It could be embarrassing if an item that is personal property was not listed because the appraiser assumed that it was real property. The appraiser may also have to perform an additional inspection of the facility to pick up the item or items that were missed.

Depending on the purpose of the appraisal, if the fair market value of the total property is requested, the appraiser considers the values and benefits that are expected to accrue as a result of the assembled entity.[6] If the assignment is to make a partial value of only certain machinery and equipment, the appraiser still makes considerations for fair market values, liquidation values, and their subcategories.

Consider the case of a large, modern, single-drum mine hoist, which meets all the current federal and state safety operating standards. First, the appraiser must determine if the hoist is personal property or real property. When appraising a building, if the building has an elevator, the appraiser is probably always correct in assuming that the elevator is real property. What makes the mine hoist, in concept, different from an elevator? Both the elevator and the mine hoist are probably used to move people and materials. If the hoist is on a substantial concrete foundation, sunk or footed 3 to 6 ft below the ground, and the hoist frame is grouted into the foundation, a case could be supported for the hoist being attached to the property.

However, if the same hoist is bolted to the foundation only, or if the hoist is on some movable framework or cribbing, the item is probably personal property. This is not a hard and fast rule. When in doubt, the appraiser can make the identification through the owner or assessor.

Occasionally a sump pump installed through the floor of a building or in a pit may be considered part of the real estate. On further observation, the appraiser may find that the pump is actually a part of the process stream, and that a pit or sump is actually an effluent catch basin.

On rare occasions, a machinery and equipment appraiser may find in a plant machinery that is used on a royalty basis. This will usually be pointed out by plant management. As an example, certain brands of

---

[6]Henry Babcock, *Appraisal Principles and Procedures*, 2d ed., American Society of Appraisers, Washington, D.C., 1980, p. 97.

packaging and sealing machinery, particularly in the fluid milk industry, are rarely owned. If the value of these machines were to be determined, it may be necessary to predict an income stream discounted to present value levels and/or to qualify the value knowing the position of ownership.

Summary

In summary, personal property usually comprises movable items, that is, items that are not permanently attached to and made part of the real estate. It is not always clear whether an item is considered personal property or real property. Court opinions are occasionally required to make the distinction between the two classes of property. The installation process can change the classification of an item of personal property into one of real property. In a majority of cases, appraisers of machinery and equipment or personal property will best serve themselves by following their instincts and common sense.

This chapter is, for the most part, limited to the distinction between two of the three major categories—real property and personal property. It is imperative that on any assignment where both the real and personal property are being appraised, the real and personal property appraisers work in unison to classify correctly the various assets into the proper categories and to ensure no duplication of values.

However, somewhere in the machinery and equipment appraiser's career, he or she will be asked to do a condemnation value appraisal for eminent domain. This should not be confusing or complicated as long as all the concepts and laws of the taking are understood.

Simply put, condemnation is for the greater good of the public (in theory), and it is the act of the sovereignty itself in place of the owner and/or the act of taking all or part of the rights of the owner.[7]

Depending on state or federal condemnation laws, there may be the consideration that the integrated components of whole properties are inseparable.[8] Where the components are merged in a complementary manner, a condemnation of the property becomes exceedingly interesting.

The argument for the integrated components of real estate/real property, personal property, and perhaps the intangibles in condemnation is

---

[7]*Real Estate Appraisal Terminology*, pp. 55, 89.

[8]Babcock, *Appraisal Principles and Procedures*, pp. 55, 354.2.

that to value only part of the entity is detrimental to the value of the whole property.[9] An example would be a pair of shoes valued at $85.00. With the loss of one of the shoes, the value of the other shoe is reduced to zero.

The following items distinguish personal property from real estate and intangibles:

1. Substance and the manner in which the item is attached to the real estate, if attached at all
2. The character of the item and its adaptation to the real estate, or the item's relevancy to the real estate
3. The intention of the party who attached the item (is the item to be affixed permanently to the property, or will it possibly be moved?)

Personal property appraisers should always be willing to admit that an item, which is considered in that jurisdiction as personal property, is outside their realm of expertise. This will only enhance and further their professionalism and standing with the client and within the profession.

---

[9]Ibid., pp. 5–6, 354.3.

3
Identification of Machinery and Equipment

Alan C. Iannacito, ASA
President, ACI Associates, Denver, Colorado

This chapter is an introduction to the identification and listing of manufacturing plants and equipment. Generally, all processes and equipment have common traits of identification. Therefore, the method of listing is not explicit to a specific industry.

Prior to the actual identification process, it is worth reviewing the purpose of the appraisal assignment. Important considerations include the scope of the appraisal work. Will the assignment require a valuation of the entire plant or is the assignment for specific portions of the plant or individual machines?

Leased Equipment

In today's manufacturing plant operations, it is likely that leased equipment is used in combination with the company's own assets. Leases may include office equipment or even an entire manufacturing line. There may be ownership of the equipment at the termination of the lease. If so, the appraiser may be asked to value the equity of the lease.

Avoid possible liability by identifying leased equipment prior to the appraisal. The appraiser's statement of *Assumptions and Limiting Conditions* should contain a clause to the effect that information furnished by the plant was accepted in good faith and to the appraiser's knowledge, leased equipment has been identified.

Appraisers have various styles of assembling information. The simplest data-gathering job can become a major project if the appraiser has not organized a procedure and defined the objective of the appraisal.

Conversely, a massive project can be simple if approached with order. Simplification is a major concern to appraiser, plant owner, and—if one is involved—lender. Simplification is no excuse for lack of detail. The other extreme in identification is overkill.

Experienced appraisers learn from training and actual work what information is important to accomplish the objective.

There are two major procedures in the identification and listing of machinery. They are identified in this chapter as macro- and microidentification.

Macroidentification

Macroidentification is a method of studying the entire manufacturing process by identifying major components contributing to the design capacity of the plant.

Interestingly enough, some components are entire plants made from a composite of items designed to work together. A prime example is an oil rig, illustrated in Figure 3.1.

Quite often we refer to an oil rig as if it is one piece of equipment. A rotary oil rig is really an entire plant made up of components. Components can be sold from the rig or replaced when they wear out. However, the oil rig is not considered complete without its complement of mast, substructures, rotary table, power source, pumps, and tools.

Some plants are a combination of plants within a plant. An entire plant may be feeding the process stream in conjunction with another part of the plant. An example of this is a plant producing feedstock for an adjacent chemical fertilizer plant.

Liquid and granular fertilizers are often made at the same location, one plant supplying the other. When the market is weak in one commodity, it may be stronger for another, causing a cutback in certain production or a shutdown. In the interim cutback, another part of the plant is capable of making product using other feedstock sources. This is demonstrated by Figure 3.2.

Identification of Machinery and Equipment 19

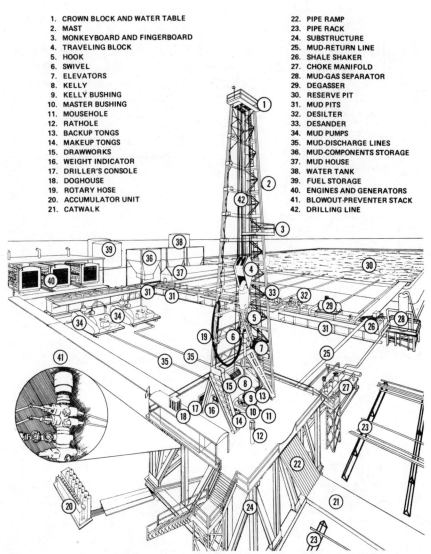

Figure 3.1. Macroidentification—an oil rig and its components. *(From* Fundamentals of Petroleum, *Petroleum Extension Service. Courtesy of the Petroleum Extension Service, The University of Texas at Austin [PETEX].)*

Using the example in Figure 3.2 we can say that *macroidentification* is the method that the appraiser uses to identify the following:

1. What the plant manufactures or produces

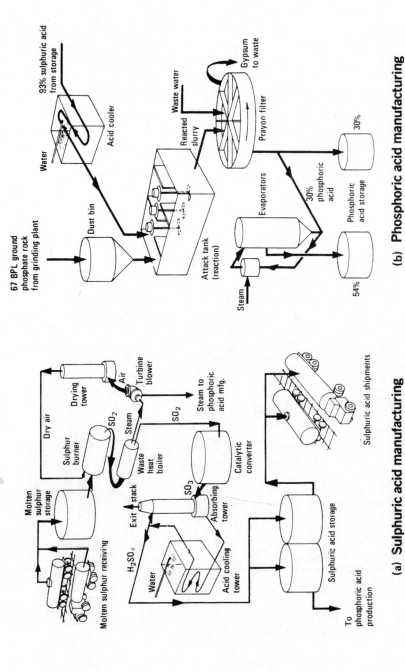

Figure 3.2. Macroidentification—example of a sulphuric acid plant. *(From E/MJ Operating Handbook of Mineral Processing, vol. 1, by R. Thomas. Courtesy of McGraw-Hill, E/MJ Mining Informational Services.)*

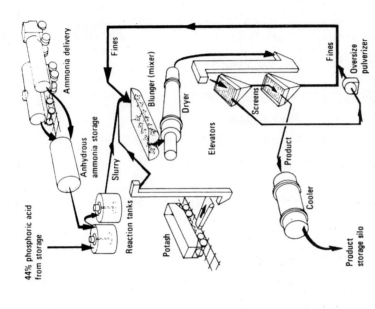

(d) Granular ammoniated phosphate manufacturing

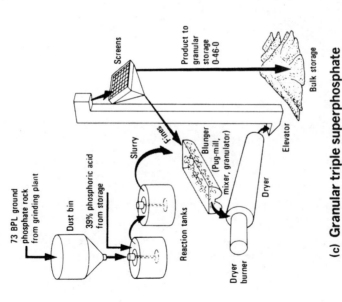

(c) Granular triple superphosphate manufacturing

Figure 3.2. (Continued)

2. How the product is manufactured
3. What the capacity of the plant is

A list of information to be considered when gathering the data for macroidentification of machinery and equipment is as follows:

1. Date
2. Company name and address
3. Who furnishes the information
4. Products produced, with each process name and description
5. Engineering design firm and contractor if other than engineering firm
6. Original date of construction and expansions
7. Plant/process by-products, amounts, and uses
8. Plant and/or unit capacity per day, tons per day, gallons per day, barrels per day, annual production, etc.
9. Plant capacities: design capacity, rated and actual consistent capacity
10. Yield or losses, reason for losses
11. Feedstocks and sources
12. Operating mode (days, month) if not identified in capacity, for example, sugar beet plants that run their "campaign only certain times of the year"
13. Outlets for finished or intermediate products
14. Plant sales outside the parent company, for use in other company plants, other products sales possibilities
15. Available historical operational data over three to five years
16. Fuel and power consumption by unit
17. Operating staff per unit: type of control systems and if the control is centralized
18. Estimated maintenance budgets over last three to five years and projected upcoming budget if plant is operational
19. Which equipment requires more than routine maintenance and why
20. How the maintenance program is conducted: regular, preventive, or demand

Identification of Machinery and Equipment 23

21. If the plant is modern and operating at total efficiency standards or if the process is obsolete, expandable, etc.
22. Plant flow: considered adequate, manageable, etc.
23. General condition of plant and components
24. Age: chronological and effective life
25. If safety and environmental standards are good (if not, can they be upgraded and at what cost)
26. Pollution control equipment in place
27. Support facilities
28. Obvious detrimental factors

Microidentification

Microidentification is the process of finding the individual characteristics of the equipment.

An example of why an appraiser will make a microidentification is demonstrated by Figure 3.3, an automated continuous ice cream mix processing line.

An appraiser may find that a similar line is part of an entire plant, similar to the separate "trains" in the fertilizer plant. The ice cream mix line is probably part of an entire dairy operation.

The homogenizer, heat exchanger, pumps, balance tanks, holding tubes, and liquefier/mixer line could be considered an entire unit. Conceivably, a replacement cost new or a liquidation value can be estimated on the entire unit.

It is likely that the equipment line will be broken up into individual components including piping/tubing, pumps, exchanger, homogenizer, wiring, and controls.

Special installation or extraordinary costs should be noted during the listing. Value-added costs in fair market valuations include foundations, electrification, plumbing, and installation. The appraiser usually does not consider value-added costs in liquidation listings.

Microidentification concerns the listing of a single machine. Identification includes the generic item. Of prime importance is the brand name, model number, serial number, type of power, and dimensions (if practical). Special accessories or controls are also included.

An example of the kind of information that is taken from the individual machine is shown in Figure 3.4.

This is a standard manually operated vertical milling machine. The sequence of information can read as follows:

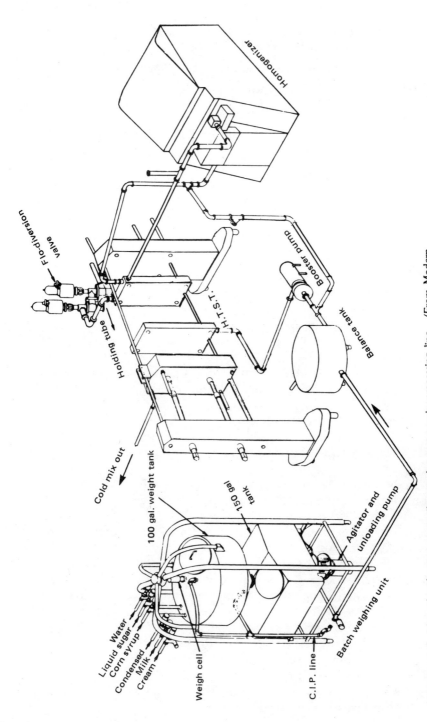

Figure 3.3. Microidentification—a continuous ice cream mix processing line. *(From Modern Dairy Products, by L. Lambert. Courtesy of America Dairy Review, Chemical Publishing Company, and the author.)*

Identification of Machinery and Equipment 25

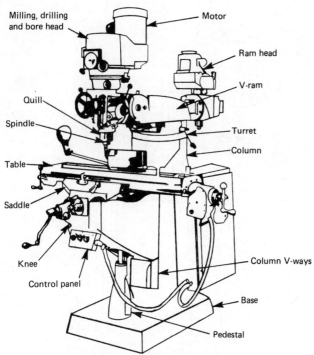

Figure 3.4. Microidentification—a vertical milling machine. *(From How to Buy Metal Working Machinery, by L. D. Slate. Courtesy of Hearst Business Media Corporation, IMN Division.)*

Vertical mill: brand name

Model number and/or size

Table size and axis (if more than two axes machine)

Special controls, readouts, programming devices

Type of milling head and motor horsepower

Accessory head, vice, tooling

Apparent condition

The time to list foundations, special installation, or other obvious value-added items is during the listings.

Another example of micro detail is the industrial fork truck demonstrated in Figure 3.5.

A typical listing reads:

Fork truck: fork lift, industrial lift truck

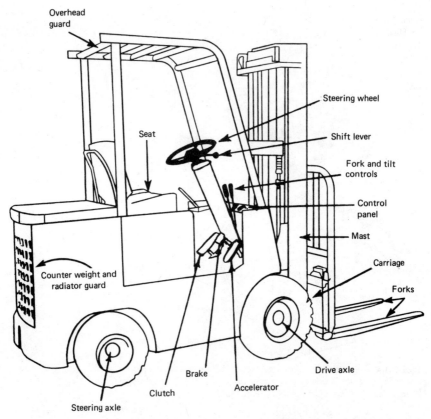

Figure 3.5. Microidentification—the principal parts of a fork truck. *(From* How to Buy Metal Working Machinery, *by L. D. Slate. Courtesy of Hearst Business Media Corporation, IMN Division.)*

Brand name and capacity

Model number, serial number

Type of tires (pneumatic, solid)

Type of truck (warehouse, rough terrain, yard)

Type of fuel (diesel, gas, propane)

If battery powered, the charger

Lift height, double mast, tri-mast

Extended fork height, tilt

Accessories: side shifter, barrel attachment, slip sheet attachment

General condition

Appendix

Following these examples, a more detailed description of what kind of information to consider in macroidentification of machines is listed as:

1. Machine type: generic name
2. Manufacturer model and serial number
3. Catalog specifications, if relevant and available
4. Size, capacity, and type
5. Materials of construction if process equipment or specialty items
6. Auxiliary equipment: special features depending on the type of equipment
7. Type of drive: V belt, chain drive, gear drive, chain to gear, reducer drive, etc.
8. Prime mover, electric motor or engine driven; if electric, the name, horse power, phase, voltage, amperage and revolutions per minute; type of enclosure; open drip proof, totally enclosed fan-cooled, enclosed nonventilated, etc.; determination if the motor is directly connected, integral to the unit (flange mounted), or connected via the drive or coupling [if engine driven: manufacturer, model, horsepower, diesel or gas; if diesel, presence of an hour meter and a scrubber for fumes; brake horsepower, at number of revolutions per minute, type of clutch, or torque converter; skid mounted, trailer mounted, permanent installation, etc.; size and type of radiator (if noted on plate); if hydraulic: pump name plate (if available), type of power, size of unit and accumulator, brand name, model and serial number; presence of excessive leakage]
9. Controls: special controls not normally furnished by the equipment manufacturer; amperage, voltages, phase, type of enclosures; if process equipment, presence of temperature recorders or other instruments relative to the equipment
10. Starting equipment, transformers, heavy duty wiring
11. Special foundations, plumbing, platforms, excessive installation costs if known or relevant
12. Other general identifying characteristics for special machinery such as construction equipment, restaurant equipment, automotive equipment, computers

It is important when recording the initial inspection that details be observed. As an appraiser's experience grows so does the ease in gathering data. Information can be refined after the appraiser has all the facts.

Appraisers are professional data gatherers. They are interested observers and good investigators. A professional appraiser is not intimidated by the lack of precise knowledge of the process or machine. If the data are good, they will be understood by novice and expert alike.

Recommended Readings

A.S.A., Boston Chapter, Machinery and Equipment Section, *Industrial Properties/ Machinery and Equipment Study Guide*, from 1976 appraisal conference, Boston, 1976.
Denver Equipment Company, *Modern Mineral Processing Flow Sheets*, 2d ed., Colorado Springs, Colorado, 1979.
Heldman, Dennis R., and R. Paul Singh, *Food Process Engineering*, 2d ed., AVI Publishing, Westport, Connecticut, 1981.
Jackson, J. M., and Shinn, B. M., *Fundamentals of Food Canning Technology*, AVI Publishing, Westport, Connecticut, 1979.
Kharbanda, O. P., *Process Plant and Equipment Cost Information*, Craftsman, Solana Beach, Calif., 1979.
Leffler, William L., *Petroleum Refining for the Non-Technical Person*, Pennwell, Tulsa, Oklahoma, 1979.
McGinnis, R. A., *Beet Sugar Technology*, 3d ed., Beet Sugar Development Foundation, Ft. Collins, Colorado, 1982.
Meade, William, P. E. (Ed.), *Encyclopedia of Chemical Process Equipment*, Reinhold, New York, 1964.
Merkel, James A., Ph.D., *Basic Engineering Principles*, 2d ed., AVI Publishing, Westport, Connecticut, 1983.
Modern Plastics Encyclopedia, McGraw-Hill, New York, yearly update, 1985–1986.
Perry, J. H., C. H. Chilton, and S. D. Kirkpatrick, *Perry's Chemical Engineer's Handbook*, 4th ed., McGraw-Hill, New York, 1963.
Potter, Norman N., Ph.D., *Food Science*, AVI Publishing, Westport, Connecticut, 1968.
Short, J. A., *Drilling, A Source Book on Oil and Gas Well Drilling from Exploration to Completion*, Pennwell, Tulsa, Oklahoma, 1983.
Slate, L. D., *How to Buy Metalworking Machinery*, Hearst Business Media Corp., IMN Division, Southfield, Michigan, 1977.
Thomas, R. (ed.), *Engineering and Mining Journal Library of Operating Handbooks*, vol. I, *Handbook of Mineral Processing*, McGraw-Hill, New York, 1977.

4
Purposes of Appraisals

David M. Graham, ASA
Greenbank, Washington

Before any appraisal can have meaning in content or significance in result, it must have a reason for being done, i.e., a purpose. A purpose is essential to establish the report asset content, the limiting conditions, or any other parameters which will lead to a proper value analysis.

A statement of appraisal purpose is a declaration by the appraiser as to the objective in preparing the valuation report. The objective may be considered from the viewpoint of the appraiser, the owner, or a third party such as a bank or a public agency. Let us look at these points of view and try to ascertain the one from which purpose can be most meaningfully stipulated.

From the appraiser's point of view, the objective is singular, namely, to establish a value for the client. Appraisers must, however, be cognizant of the intended use of the appraisal so that they may employ in their value analyses the proper theoretical and practical considerations that will lead to a valid conclusion. While the appraiser may choose the particular type of value (e.g., fair market, insurable) to be presented and the methods of study employed, these choices must be guided by the knowledge of how the appraisal will be utilized.

Statement of Purpose

Accordingly, an appraiser might state in the report, "The purpose of this appraisal is to express an opinion of the fair market value of the subject assets." This may sound just fine and may even be the limit of commitment a client wishes expressed in a formal report. It is basically, however, an inadequate or incomplete statement of purpose because there are many reasons why a fair market value might be needed. It would be far more definitive to state "The purpose of this appraisal is to express an opinion of the fair market value of the subject assets to serve in purchase/sale considerations." Such a specification of purpose would not only be apropos but also would provide a safeguard against possible misuse of the appraisal.

This leads to the conclusion that the basic reason to do an appraisal is to meet an owner's business or personal need for a statement of asset values that is appropriate for use related to that need. The owner's need to accomplish a specific objective therefore creates the genuine and legitimate purpose for having an appraisal done. The owner may need a determination of asset values because he or she wishes to acquire adequate fire insurance coverage, borrow money for the purchase of new equipment, negotiate a merger with a company in an allied industry, or substantiate his or her position in a condemnation action.

Third Party Requirements for an Appraisal

It is realistic to note, however, that an owner's need for an appraisal may be the result of the requirements of a third party. In fact, the third party may be the client. Some examples are as follows:

> Insurance companies want to know the size of risk they are underwriting or verify compliance with a coinsurance clause.
>
> Lending institutions want to know how much they can safely loan and expect to recover in the event of default.
>
> Public agencies want to know a fair offer to make for acquisition for public use.

Realizing that an appraiser's client may be an owner or a third party, we can say that the purpose of any appraisal is to facilitate the endeavors as contemplated by one of the following:

Accountants	Condemnees	Lessees
Agents	Condemners	Lessors
Assessors	Corporations	Managers
Attorneys	Courts	Owners
Brokers	Creditors	Partners
Buyers	Insurers	Sellers
Commissioners	Lenders	Spouses

Multiple Purpose Appraisals

One important consideration, however, is that on occasion an owner may have more than one purpose for having an appraisal done. For example, the owner may need to establish an insurable value to acquire adequate fire insurance coverage and liquidation value to guarantee a loan from the bank for operating capital.

While these both are valid purposes, the value analysis is quite different for each purpose. Field inspection, inventory listing, and replacement cost research may be the same for both appraisals. The final value analyses to derive appropriate values to serve each purpose properly, however, are quite different.

Depending on the wishes of the owner, the end results may be presented in one report, in separately headed columns, or in two completely separate reports so that ultimate users are provided information only on a need to know basis.

We must not overlook the fact that often appraisals are used in or by a court. It is possible, though rare, that a court is the client. Most likely court attention to an appraisal is in litigation which has resulted from the failure to resolve some value question otherwise.

It is extremely important to note that many machinery and equipment appraisals may be fractional appraisals, particularly with respect to those done for any purpose which requires the fair market value determination. Since many appraisals can and often do include a scope beyond machinery and equipment, the appraiser must be careful to define clearly any limiting conditions and/or the fractional character of the report.

Relationship of Purpose to Value Concept

Table 4.1 identifies the various purposes for which an appraisal might be required and the value concept germane to that purpose. Following

Table 4.1. Value Concepts for Appraisal Purposes

Appraisal purpose	Replacement cost new	Fair market value	Liquidation value	Scrap/ salvage	Insurable value
Allocation of purchase price		X			
Bankruptcy			X		
Business valuation		X			
Condemnation		X			
Cost studies	X*				
Dissolutions		X*			
Estate planning		X*			
Incorporation		X			
Insurance					X
Insurance loss settlement					X
Loan			X		
Management considerations		X*			
Merger		X			
Partnership formation		X			
Purchase/sale		X*			
Stock issue		X			
Surplus disposition				X	
Taxation—ad valorem		X			
Taxation—gift, estate		X			

*Principal concept—others might be appropriate.

the table are brief discussions of each purpose listed in Table 4.1. Value concepts are the subjects of other chapters.

Allocation of Purchase Price

Often entities containing multiple assets are purchased as a unit. Subsequently, there may be a need to know the values of the separate components of the purchase in order to set up appropriate asset records, depreciation schedules, or similar information. A fair market value study becomes necessary to establish a proper proportioning of the acquisition cost against the various components.

Bankruptcy

An appraisal done for the purpose of bankruptcy considerations is a *liquidation value*. The importance of this analysis is that creditors usually prefer to be paid in cash, not equipment assets. A liquidation value appraisal should therefore estimate the likely net recovery from the forced sale of the assets. In all respects, this analysis follows the same pattern of logic as for loan purposes, since in essence creditors are the same as lenders.

Business Valuation

Sometimes the purpose of having an appraisal done is to establish a business valuation. Of itself, this purpose does not sound like it accomplishes anything other than to satisfy the desire to know on the part of an owner. Indeed, that may well be the only purpose.

It is more than likely, however, that some more specific purpose may be intended, such as sale, merger, or incorporation, wherein a fair market value analysis is in order. For whatever reasons, however, the owner may wish the appraisal to be rather nonspecific in statement of purpose. The owner may, in fact, have two or more purposes in mind, though all purposes may be properly served by the fair market value analysis.

It should be noted that a business valuation is all-inclusive of tangible and intangible assets and, in total, may have a scope well beyond the appraisal of machinery and equipment.

Condemnation

When a public agency needs to acquire private property for conversion to public facilities, it must exercise the right of eminent domain. Often the property in question is industrial or commercial in character and may contain extensive machinery and equipment. When such a project occurs, both condemner and condemnee have a compelling need to obtain an appraisal as a basis for negotiation or for court presentation should litigation follow unsuccessful negotiation. The value concept involved is mandated by law as the fair market value.

Cost Studies

Cost studies are the only appraisal purpose for which the determination of *replacement cost new* is the end result. Such a study would be of primary use if a client were analyzing the feasibility or alternatives of plant expansion, change, modernization, or relocation.

A *cost study,* i.e., the determination of replacement cost new, is fundamental to the establishment of insurance, which may be written on a replacement form. Obviously, the same is true for an insurance loss settlement related to insurance written on a replacement form. In these cases, however, the cost study is not itself the real purpose.

It is most important to realize that the consideration of replacement cost new (cost study) may be required in many appraisals, regardless of their otherwise stated purpose, as a starting point in the *cost approach to value*, i.e., the cost from which depreciation and obsolescence are deducted. Here again, however, the cost study is not, per se, the actual purpose of appraisal.

It should be noted that a cost study could be, in fact, a study of costs on a used market basis.

Dissolutions

Appraisals for the purpose of dissolutions, whether they be marriages, partnerships, or corporations, are usually done on a fair market value basis in order to establish an equitable distribution to each spouse, partner, or stockholder.

It is possible that an appraisal for dissolution could be on an estimated liquidation value basis, particularly if the involved parties wish to consider disposition of all assets first and then division of the monetary results.

Estate Planning

Appraisals of assets on a fair market value and/or other value basis might be required by an individual involved in estate planning. This is because estate planning may include consideration of a variety of assets and multiple objectives such as specific bequests, tax avoidance, or trust formation, for example.

Incorporation

Each state has a body of laws, administered by public authority, which establishes the requirements under which a corporation will be permitted to exist and operate. An *incorporation procedure* is a registration with that public authority of all relevant information necessary to establish a corporate entity. This includes a statement of worth and may require an impartial professional appraisal of all assets on a fair market value basis.

The formation of a corporation implies the issuance of stock. Accordingly, the value of the corporation must be determined in order that a per-share value of stock may be established. The subsequent sale of stock may be public or private, but it will be subject to the rules of the controlling public authority.

Insurance

The determination of an insurable value is one of the most common purposes for which appraisals may be required. The need here is to establish a value which represents an adequate amount of insurance to purchase to indemnify the insured against loss. Such a value is of concern to owners, lessors, or lessees (for liabilities), insurers, agents, and brokers.

The insurable value may be determined by a replacement cost new if insurance coverage is provided on a replacement form. More often, however, the insurable value will be predicated on replacement cost new less depreciation or (the comparables approach) a cost used of like kind, quality, and condition plus appropriate freight, taxes, and installation. This is often referred to as the *actual cash value*.

Since many insurance policies covering industrial and commercial machinery and equipment include a coinsurance (average) clause, a good appraisal is essential to forestall the possibility of failure to collect in the event of loss.

Insurance Loss Settlement

An appraisal done for an insurance loss settlement has a very special and limited purpose, i.e., to verify that asset values are in compliance with insurance policy requirements. The values determined are the same as for the insurance appraisal. The only real difference is that for loss settlement the appraisal is done after the loss has occurred.

Loan

An appraisal for loan purposes is done by or for a lender. Here the primary concern is what is the value guaranteeing the loan. In other words, if a borrower defaults on loan payments and the lender must foreclose, the value is what can be recovered from the distress sale of the assets pledged to guarantee the loan. Lenders do not want the assets. They want cash, and as soon as possible. It is therefore obvious that a liquidation value is the concept of value relevant to an appraisal for loan purposes since a distress sale will generate cash quickly.

Management Considerations

This is a catchall appraisal purpose that may sometimes be called *internal financial considerations*. It is usually required by owners only. The real appraisal purpose may be one or more of the reasons why a fair market value and/or any other analysis is appropriate. The management considerations purpose may even be a business valuation, or it may be something less, wherein the report content excludes intangibles, cash assets, real estate, or whatever else is specified. As in business valuation, however, the owner may wish the appraisal statement of purpose to be rather nonspecific.

Merger

A *merger* is a joining of two entities. It may be considered that two separate entities are each selling to the other, such that each has an appropriate proportional ownership interest in the resulting single entity. These interests are properly determined by fair market value analyses of the assets each contributes to the union.

Corporations are the common entities to join in a merger, and a single line of authority results from the original two. Usually, the identity of the dominant entity is retained, while the minority interest either loses its identity or becomes a division of the dominant entity.

Partnership Formation

This purpose of appraisal is basically the same as in a merger involving two or more individuals or companies. Unlike in merger, however, and while there may be a distinction between senior and junior partners, the partners are joint principals and retain their individual identities. The interests are determined by fair market value analyses.

A *joint venture* is a partnership of individuals or companies who have joined together for a specific project which is limited in time, operation, or both.

Purchase/Sale

One of the more common purposes of providing a fair market value appraisal is to establish a basis upon which a willing buyer and willing seller might negotiate. The need for appraisal is likely of equal importance to both buyer and seller.

It is not unusual for a buyer and seller to agree in advance to be bound by the fair market value determination of a single appraiser and split the appraisal fee.

Stock Issue

An appraisal done for the purpose of issuing stock should be on a fair market value basis, since it is really a form of purchase/sale. The stock purchaser is buying and the stock issuer is selling some percentage of ownership.

This purpose might be considered the same as, or at least concurrent with, an incorporation procedure. It may also be relevant to a subsequent stock issue by an existing corporation.

Surplus Disposition

This purpose of appraisal is infrequently required of professional appraisers. The need is more likely served by an owner's direct negotiation with a scrap dealer, and it is seldom associated with any proportionately large monetary considerations. The appraisal need may result from a client's desire to clean out a warehouse or dispose of a "bone pile" in order to create usable space. The value concept involved is scrap/salvage. For this reason, *surplus* should not include good, usable machinery and equipment for which there may be a demand on the used market and which might be more properly and profitably considered under the value concepts of fair market value and/or liquidation value.

Taxation—Ad Valorem

Appraisals for this purpose are to establish asset worth on a fair market value basis for property taxes to be assessed. Such appraisals are made by a tax assessor. However, property owners who believe they may have been assessed unfairly and wish to appeal may have their own appraiser determine values as a basis upon which to contest the assessment.

Taxation—Gift, Estate

Appraisals made for the purpose of establishing a gift or estate taxable basis are also done on a fair market value basis by appraisers appointed to such service. As with taxation—ad valorem, owners may wish to appeal an evaluation based upon an appraisal by their own representative.

Each appraiser may have a preferred wording for the statement of appraisal purpose as recorded in his or her reports. Accordingly, the statements of purpose that follow are not intended to encourage expressive conformity of all appraisers but rather to illustrate inherent differences in meaning.

> The purpose of this appraisal is to express an opinion of the fair market value of the subject assets to serve in purchase/sale considerations.

This is the most definitive statement in that both the intended use of the appraisal and the value concept employed are specified. It should, therefore, be considered the preferred form of statement of purpose.

> The purpose of this appraisal is to express an opinion of the fair market value of the subject assets to serve in management considerations.

This statement of purpose should be considered second choice because, while it specifies the value concept employed, the intended use of the appraisal is rather nonspecific.

> The purpose of this appraisal is to express an opinion of the fair market value of the subject assets.

This statement of purpose is third choice for inclusion in an appraisal, since no purpose of use is specified. If the client so insists, however, the appraiser may be obligated to record only this limited, albeit acceptable, statement of appraisal purpose.

5
Replacement Cost New Concepts

Merritt Agabian, ASA
President, A&M Appraisal Company, East Walpole, Massachusetts

The starting point of an appraisal using the cost approach is the replacement cost new, installed. The theory is that an item is worth no more than the replacement cost new, installed and indeed may be worth less because of its condition. Thus, the *replacement cost new, installed* is the upper limit of value in a machinery and equipment appraisal.

Definition of Replacement Cost New

The Machinery and Equipment Committee of the American Society of Appraisers defines the *replacement cost new* as the current cost of a similar new item having the nearest equivalent utility as the item being appraised.

How is this replacement cost new, installed, obtained? If a machine is still being built, it is the current cost of that new machine. If the subject machine is no longer in production, the superseded machine may qualify as the replacement machine. If the manufacturer is no longer in business, it may mean the replacement by another manufacturer.

Sources of Replacement Cost

The replacement may be for an individual machine or for an entire plant. The replacement cost new for an entire plant or process can be estimated using cost-to-capacity factors. Some of these factors are found in cost manuals, and others can be established by the appraiser from publications and public records. This type of information is useful in situations where the appraiser is acting as a consultant but should always be followed up with a full appraisal if more dependable costs are required.

Care should be taken to modify the factors for size. A 100 ton per day plant that costs $100,000 per ton does not mean that a 500 ton per day plant would cost 500 × $100,000 = $50,000,000. There are economies of scale that must be considered. The so-called six-tenths factor of C. H. Chilton[1] refers to this economy of size relationship for chemical plants. The American Association of Cost Engineers has made refinements to this power rule as it applies to various processes.

Definition of Reproduction Cost New

The *reproduction cost new* is the current cost of duplicating an identical new item. This is the Machinery and Equipment Committee definition. Seldom is a machine duplicated in practice. Even current modern machines are improved constantly in the manufacturing process. The machine may be completely redesigned, new materials specified, and/or drive arrangements and horsepower changed. Even minor items such as individual bearing shafts or gears may be improved or changed at any time. Unless the changes are of a major consequence, the machinery appraiser may not even be aware of many routine improvements or changes in a machine. If the manufacturer still calls it a "Model E," we

---

[1]C. H. Chilton, "Six Tenths Factor Applied to Complete Plant Costs," *Chemical Engineering*, April 1950. Copyright ©1950 by McGraw-Hill, Inc., New York, N.Y. 10020.

assume that it is a replica of our five-year-old Model E machine and we could call the current price the reproduction cost new.

Determining Replacement Cost New

Inquiry should be made if the manufacturer changes the designation to "Model E-1" and even more so when the new machine is designated as a "Model F." What changes have been made? Is the Model F faster or more productive than the E? Does it have greater capacity or better tolerances? Does it have features that make it more durable or flexible? Possibly the current cost of the Model F would be the replacement cost of the Model E.

So, for a current modern machine that is still manufactured with the same model designation, it could be said that the cost arrived at is a reproduction cost new. It is also the replacement cost new.

When the item being appraised is an older machine, the process becomes more complicated. No one builds a replica of a 1920s era cone head lathe or a flat belt-driven milling machine. The cost to secure patterns for castings, tooling, and manufacturing would result in a cost probably in excess of a *new* similar size lathe or milling machine. Surely, then, the reproduction cost new is not the cost that is being sought.

The replacement cost new would have more meaning, but the replacement lathe and milling machine are all geared with more feeds and speeds and perhaps even more rigidity and power for much higher tolerances and production of work. To properly evaluate the older machine, these improvements must be assessed and adjustments made to the replacement cost new, either in the new cost column or through depreciation in the value column.

Some appraisers make the adjustments in the New Cost column for two reasons:

1. The replacement machine is so much superior to the old machine that its use in such a situation is misleading.
2. For insurance appraisals to replace the old machine with a new machine (in repair and replace endorsement) is putting the insured in a better position than before the loss.

Special Purpose Machines

Occasionally an appraiser encounters a situation in which a machine is specially built for a specific operation. The machine is unique and operates to the complete satisfaction of the owner. The only alternative is to build a new machine from the same plans. In such a case, the *reproduction cost new* is the proper term to use.

In other instances through technological improvements, the replacement machine may have all the utility and features of the subject machine but sell for less money than the subject machine cost several years previously. An example of this is the computer industry where performance has improved and prices have been reduced. The *replacement cost new* for similar capacity is the cost to use in this instance.

However, in a machinery and equipment appraisal many items are encountered. In some instances, the *replacement cost new* is the more accurate title; in other instances, the *reproduction cost new* is the more appropriate heading. To use both terms in an appraisal is cumbersome and creates problems. Many machinery and equipment appraisers use the terms interchangeably[2] or use one heading in the listing of equipment and indicate with an asterisk when another basis is used. For example, the replacement cost of an old mechanical calculator could be used as opposed to the reproduction cost new of an old mechanical calculator.

In practice some appraisers use replacement cost new and others use reproduction cost new. Experienced appraisers have encountered these factors and have resolved them mentally and should also explain them in their reports to clients so that there is no confusion concerning the costs that are reported.

Information Required

To determine the proper replacement cost new, the appraiser must secure the right information during the inspection.

Essential data include the manufacturer, model, serial number, capacity, attachments, material composition, type and size of the drive, and type and extent of the installation.[3] These items are used in their

---

[2]George D. Sinclair, "Appraisal Concepts," *The Appraisal of Machinery and Equipment*, ASA Monograph no. 2, September 1969, p. 13. Used by permission of American Society of Appraisers.

[3]Thomas L. Schropp, "Identification," *The Appraisal of Machinery and Equipment*, ASA Monograph no. 2, September 1969, p. 33. Used by permission of American Society of Appraisers.

order of importance in correctly costing the machine. Without the manufacturer, you have to rely on generic information. We all know that both the Ford Escort automobile and the Rolls Royce Silver Shadow automobile are sedans, but there is an enormous difference in cost. The same applies to presses (an Alva Allen press versus a Minster press), drills (an Enco drill versus a Rockwell drill versus an Allen drill), lathes (an Atlas lathe versus a Hardinge lathe), and all types of machine tools and other items of equipment and plant furniture. The cost difference of large forging hammers of two American manufacturers (16,000 to 50,000 lb) varies by 40 percent. This is surprising as both are quality firms and the design of drop hammers has not changed materially in many years. Upon examination, the more expensive firm's hammers weighed 30 percent more than the other company's hammers. That goes a long way toward explaining the price spread.

Most firms require the model designation to quote a current selling price. Likewise, the serial number is the key for the manufacturer to identify a machine, know when it was built, what attachments it has, whom it was sold to, and even the present and past owners. Some firms will not quote information and prices without a serial number.

Sometimes the model number in itself is insufficient for securing the correct cost. The capacity is an important consideration on almost every item of equipment. The swing and center distance of a lathe, the milling length and width of a milling machine, the swing of a radial drill, the number of spindles of a drill, the size of bar stock on an automatic, and the tonnage of a press are all items of capacity.

The installation factors are also important in the determination of the replacement cost new. Does the machine have a special foundation, pilings, long utility runs, difficult ingress? These and other items of installation must be noted to arrive at the proper cost.

The best source of information for the machinery and equipment appraiser is the manufacturer or dealer of that particular item. Inquiry to the manufacturer establishes whether the machine is still in production (being made), its present availability or time necessary for manufacturer and delivery, its weight (to help estimate shipping or freight charges), and especially its current cost new and costs of attachments and accessories. If the machine is no longer in production, the manufacturer can often supply the cost of a machine that replaces the subject machine. The fact that a machine is no longer being made in itself gives some indication of the value of that machine. Older machines may not have replacement parts available and so are much more costly to repair.

Cost Data Procedures

Cost manuals, catalogs, and price lists are a valuable source of cost information.[4] These sources are dealt with in Chapter 6. Every machinery and equipment appraiser should establish and maintain a cost library for accurate information and for the production efficiency it fosters.

The trending of original costs has become a popular method of establishing the replacement cost new with many appraisers. Often the asset listing of a subject company is available given the original cost and date of acquisition. Applying a cost trend published by various cost services and appraisal and insurance companies is a fast way to estimate the replacement or reproduction cost new. The result is either replacement cost new if replacement items are included in the makeup of the indices or reproduction cost new if costs are adjusted to account for technological improvements, as is usually the case.

Trending Cost Data

The trend index is usually for some category of equipment or for a particular industry. The time intervals are indicated with a corresponding numerical amount representing an average cost level at that particular date. The trend index for machine tools may look like Table 5.1. It can be used to determine cost with the following equation:

$$\frac{\text{Current year index}}{\text{Known year index}} = \frac{325 \ (1985)}{265 \ (1980)} = 1.226, \text{ say } 1.23$$

With a machine known to cost $10,000 in 1980 and a current cost is required, the 1980 multiplier of 1.23 would be applied to $10,000 to arrive at $12,300, the cost in 1985. Of course, the index indicates relative cost levels and can be used to trend a current cost to some previous date as may be required by an appraisal assignment.

Trending is fraught with potential pitfalls. First, is the original cost the actual first cost or was it an allocation made during some prior transfer of the property? Second, what items were included in the original cost? Were all installation charges included? Third, was the original cost for an item that was bought used? Surely, trending a used cost will not produce a replacement or reproduction cost new. Finally, how good

---

[4]Henry A. Garber, "Sources of Data," *The Appraisal of Machinery and Equipment,* ASA Monograph no. 2, September 1969, p. 67. Used by permission of American Society of Appraisers.

is the index trend you are using and, more importantly, does it Many trends are made up of a market basket of items in a particular category. The more nearly your assets mirror the assets in the market basket, the more likely you are to achieve a reasonable result. Quite often we do not know how a trend index is formulated. The category headings are vague or very general. At any rate using a trend factor for periods in excess of 10 years should be done with extreme caution.[5]

Another method that may be preferable to straight trending is the trending of the items of a manufacturer. If a particular machine increased 25 percent since 1980, another machine of the same manufacturer probably increased by approximately 25 percent in the same time frame. Sometimes you can keep cost index trends for a particular manufacturer. Again, care should be exercised. A manufacturer raises prices on items that are in demand and that sell readily. He or she is reluctant to raise the price of an item that is a slow mover even in times of inflation.

Trending does have its place in the appraisal profession, but the wholesale trending of all costs to estimate the replacement or reproduction cost new and to serve as a basis of fair market value is dangerous and could bring about the end of the appraisal of machinery and equipment as we know it.

Table 5.1. Example of a Trend Index for Machine Tools

Year	Trend index	Multiplier
1985	325	1.00
1984	318	1.02
1983	315	1.03
1982	312	1.04
1981	295	1.10
1980	265	1.23
1979	248	1.31
1978	216	1.50
1977	200	1.63
1976	191	1.70
1975	180	1.81

[5]Nobel L. Davis, "Machinery and Equipment Trends—How Are They Used?," *Valuation*, vol. 17, no. 2, 1970, p. 64. Used by permission of American Society of Appraisers.

Other Methods for Determining Costs

Comparing the subject asset with an item manufactured by another company is risky, but in some circumstances it must be done. One such occasion is when the subject manufacturer is no longer in business or does not produce that particular line of equipment.

Estimating the weight of a machine and applying a price per pound is sometimes used, but again it is probably the least desirable method of estimating the replacement or reproduction cost new. A direct estimate of the cost of a machine is quite often just as accurate as estimating weight and should be done when it is anticipated that obtaining cost information will be difficult or impossible.

Occasionally we encounter an industry where advancements in technology, design, or materials make possible the production of a machine that costs a good deal less to manufacture. For instance, if a machine produces 1000 tongue depressors per hour and costs $10,000 to manufacture and replaces two machines that produce 500 tongue depressors per hour each, the replacement cost new for the 500-per-hour machine would be $5000. Care should be taken in using this approach because the company may lose some production flexibility with fewer machines. Also, if the 500-per-hour machine is capable of producing all the items needed by that company, the reproduction cost new must be the full cost of reproducing that machine.

Components of Replacement Cost New

Various components make up the replacement-reproduction cost new. These are:

1. Purchase price new of the machine with all its features and attachments (crating and skidding may be required)
2. Freight—the total cost to load, ship, and unload at the purchaser's dock
3. Rigging—either by plant personnel or by outside contractors
4. Erection or assembly when a machine is shipped in parts
5. Installation, including foundations, pilings, utility connection, trial runs, and debugging
6. Indirect costs and fees for machine evaluation for purchase selection, plant layout, necessary licensing fees, and taxes

The extent of all these factors varies from machine to machine and location to location. Machinery and equipment appraisers must familiarize themselves with these components and methods to estimate their cost.[6]

We have discussed the purchase price new and how to secure this cost or how it can be estimated.

The *freight charge* can be determined by contacting railroads, trucking companies, and, in some instances, shipping and air freight companies. The type of equipment that is to be shipped must be known as well as the weight of the item.

The rigging costs can be determined by estimating the crew time and the equipment needed for the job. Again, consulting with a rigging company can provide guidelines in making the estimate more accurate.

The manufacturer is probably the best source for information regarding erection or assembly time and costs.

Installation charges are varied and can involve several trades. The foundation costs involve carpenters for forms, cement finishers, operators for pile drivers, steamfitters, laborers, bricklayers, millwrights, welders, and workers from any of the other building construction trades if a building must be opened to accommodate the installation.

Ordinary trial run-in costs and adjustments are legitimate additions to cost, but prolonged debugging charges should be avoided.

The indirect charges for design, engineering, supervision, licensing, permits, and taxes must be considered. In some process industries, these indirect charges can exceed the direct charges. Some items of equipment require a license to operate. The costs of acquiring a Federal Communications Commission license for a VHF television station can far exceed the costs of the equipment involved, and its value usually exceeds the cost and is usually a high multiple of the equipment costs. The value of such a license could be better estimated by income and business valuation techniques.

Production equipment is exempt from sales taxes in most states, but the appraiser must determine that fact and what operations qualify to have production equipment. Usually service equipment and office furniture require the payment of a sales tax.

On appraisals of large industrial plants, most appraisal companies have formulas or schedules to estimate the various installation factors more rapidly. Some of these techniques are to add a percentage amount to the selling price new of a machine or assign a dollar amount per

---

[6]Thomas L. Roberts, "Machinery and Equipment in Case of Eminent Domain," *The Appraisal of Machinery and Equipment,* ASA Monograph no. 2, September 1969, p. 81. Used by permission of American Society of Appraisers.

pound of a machine. These can be used after the appraiser has had some experience with estimating the costs on an individual component basis. Where the replacement or reproduction cost new is the basis of establishing the fair market value, in those instances in which there can be litigation the appraiser would be well served to individually estimate the components of installation.

Cost Versus Value

One final word: the proper terms to use are replacement *cost* new or reproduction *cost* new and not replacement *value* and reproduction *value*. Webster defines *cost* as "the amount or equivalent price given or charged or engaged to be paid or given for anything." It is a fixed amount that is charged for an item by a seller and presumably any buyer would pay the same price. It can be verified. It is the exact amount a company actually pays to acquire an asset.

Value is usually estimated and is subject to differences of opinion. It is defined in Webster as "a fair return in money, goods, services, etc., for something in exchange; monetary worth of a thing; marketable price." It is subject to interpretation and varies from one person to another. As stated in at least one other chapter in this book, value is what a thing is worth; price is what is paid to acquire it. Using the term replacement *value* is not recommended and indicates that the amount is subject to various opinions.

Conclusions

This chapter is intended to highlight the concept of replacement cost new as it pertains to the appraisal process. The differences in replacement cost and reproduction cost are explored, and the components of these costs are examined.

In some appraisals the replacement or reproduction cost new is the result that is required. More often it is the beginning point in an estimate of the fair market value. The vehicle for extracting the fair market value from the replacement cost new is depreciation, which is discussed in Chapter 7.

6
Sources of Pricing and Reference Material

Paul Rice, ASA
Manager, Appraisal and Valuation Services, Arthur Andersen & Company, Los Angeles, California

The purpose of this chapter is to give the entry-level appraiser an insight into the sources of information needed to derive an opinion of value for machinery and equipment under appraisement.

All appraisers, regardless of their professional discipline, must do research on pricing and/or reference materials. The machinery and equipment appraiser is fortunate because there is a wealth of pricing and reference materials available, e.g., manufacturers' and dealers' catalogs, price lists, and equipment specifications.

Pricing References Are Important for Machinery and Equipment Appraising

Every machinery and equipment appraiser must have a readily available pricing and reference library. This information, a systematic collection of current data, is a "must" to ensure that opinions of value are valid and supportable. Support data are the backbone of the appraisal and must be readily available should justification be needed in the event of litigation or governmental hearing. The more complete the pricing and reference material library, the better able the appraiser will be to support his or her findings.

How the pricing and reference material is set up will be up to your personal preference. Most libraries are set up alphabetically, by manufacturer, by coded numbering system, basic industry classifications, or combinations of any of these systems.

Basic Source Materials

A good pricing and reference material library should contain as much relevant information as can be reasonably obtained, such as:

- Manufacturers' specifications manuals
- Manufacturers' catalogs
- Manufacturers' price lists
- Vendors' price lists
- Published price guides
- Serial number reference guide
- Office of Price Administration machine tool base prices (1953)
- Reference books

Pricing and reference materials can be obtained from the following sources:

- Manufacturers and vendors
- New and used machinery dealers
- Trade shows and exhibitions
- Published price guides
- Client invoices
- Newspaper classified sections
- Auctioneers
- Universities
- Public libraries

Computerized data bases

Previously completed appraisal reports (in house)

Approaches to Data Sources

Now that you know what you need and where to acquire your pricing and reference materials, the next step is to know how to approach the manufacturers, dealers, and fellow appraisers. Contacts can be made by:

Telephone (see Figure 6.1)

Letter (see Figure 6.2)

Personal contact

Whatever method is used, it is necessary to build rapport with the individuals and companies with whom you are dealing. Let them know how much you appreciate their help, verbally and through written communication. Remember, everyone likes to be appreciated. In addition, let them know that you will advise your client (their customer) of their cooperation.

If you develop the proper rapport with manufacturers and dealers, you will not only acquire price and reference material, but you may also be apprised of economic trends, present market conditions, and anticipated future developments. These will help you to make the proper evaluations regarding economic and functional obsolescence factors.

Personal contact with other appraisers could be one of the greatest sources of pricing and reference materials. Remember, they are in the same position as you, and not everyone can be an expert or secure in-

---

NAME OF MANUFACTURER:

ITEM:_____
REMARKS:_____
DATE:_____
Local Phone #:_____
Serial #:_____
Model#:_____
Weight:_____
FOB:_____
Base Price: $_____
Excise Tax: $_____
Total: $_____

Figure 6.1. Telephone price quotation form for machinery and equipment appraisal.

DATE:_____
  TO:
  RE: (Client)
  Gentlemen:
  Please furnish current consumer prices for the following items listed below. This information will be considered as confidential and is required to establish appraisal values for our mutual client.
  Please note that we desire only the present selling prices to this client and that the cost of the original equipment is not desired. In case an item is obsolete, please quote on its replacement and indicate description and model number.

  <u>Item</u>   <u>Weight</u>   <u>Price FOB</u>
  (To be completed by appraiser)

  Thank you for this information.
  Very truly yours,

Figure 6.2. Letter request for price quotation.

formation on every subject. Therefore, we should be willing to share our information when the need arises. Over the years, this has not generally been done. Many reasons are readily apparent, but, probably, the most important has been the competition for appraisal engagements among individuals and companies. We must somehow overcome this problem because cooperation can be rewarding and beneficial to both parties. Sharing pricing and reference material does not mean giving away trade secrets or turning over your entire library; it is a way of building a firm data base. The electronic data base is being used on a limited basis. As more work is completed, this may become a significant pricing and reference tool.

Following is a listing of sample publications and recommended readings. We must caution the reader that this list names only a small percentage of the publications and readings available in today's market. It is recommended that you continually add to these lists and share them with other appraisers.

Miscellaneous Free Publications (New and Used Equipment)

Construction Equipment Advertiser

Local newspapers

Machinery Journal tool catalog
Metal Fabricating News
Print-Equipment News
Truck Gazette
Yellow Pages telephone directory

Manufacturer's Catalogs

Big Book of Metal Working Machinery (no longer in print, but may be found in used bookstores)

Blackhawk (hand tools)

Brodhead–Garrett Co.

Hewlett-Packard Co.

John Fluke Mfg. Co., Inc.

Koch Supplies, Inc.

McMaster Carr Supply Co.

Owatonna Tool, Inc.

Rotunda (Ford) Automobile Equipment

Sun Electric Corp.

Tektronix, Inc., Instruments Group

There are many more publications available, but the above are representative of the kinds of information successful machinery and equipment appraisers should have in their files.

Published Price Guides (Prices Vary)

Aircraft Blue Book, Aircraft Blue Book Corporation, Will Rogers World Airport, PO Box 59977, Oklahoma City, OK 73159

Blue Book—Auto, Kelly Market Report, 2950-A7 Airway Avenue, Costa Mesa, CA 92626

Blue Book of Current Market Prices of Used Heavy Construction Equipment, Forke Brothers, 830 NBC Center, Lincoln, NE 68508

Building Construction Cost Data, R. S. Means & Co., Inc., 100 Construction Plaza, Kingston, MA 02364

Computer Hardware & Software, Data Services, PO Box 5845, Cherry Hill, NJ 08034

Computer Price Guide, The Blue Book of used IBM Computer Prices, 75 South Greely Ave., Chappaqua, NY 10514

Dictionary of Scientific and Technological Terms, McGraw-Hill Book Company, New York, NY 10020

Directory of Industry Data Sources, Ballinger Publishing Company, Division of Harper & Row, Cambridge, MA 02138

Energy Cost Reference Book, AACE, 308 Monongahela Building, Morgantown, WV 26505

Equipment Directory of Audio Visual Computer and Video Products, International Communications Industries Association, 3150 Spring Street, Fairfax, VA 22031

Food Processors Guide, Food Processing Machinery and Supplies Association, 1828 L Lane NW, Washington, DC 20036

Green Guide for Lift Trucks, Data Quest, Inc., 1290 Ridder Park Drive, San Jose, CA 95131

IMN Auction Report, Hearst Business Media Corporation, 29516 Southfield Road, Southfield, MI 48086

Industrial Machinery News, 29516 Southfield Road, PO Box 5002, Southfield, MI 48037

Marshall Valuation Service, Marshall & Swift, 1617 Beverly Boulevard, PO Box 26307, Los Angeles, CA 90026

Modern Cost Engineering Methods and Data, vols. I and II, *Chemical Engineering,* 1221 Avenue of the Americas, New York, NY 10020

Nomda's Blue Book Industry Guide to Market Prices & Manufacturers' Trade-In Schedules, National Office Machine Dealers Association, PO Box 707, Wooddale, IL 90191

Official Guide—Tractors and Farm Equipment, Southwest Hardware and Implement Association, 4629 Mark IV Parkway, Fort Worth, TX 76106

Richardson Engineering Services, Inc.—Process Plant Construction Estimating Standards, 4 vols., 909 Rancheros Drive, PO Box 1055, San Marcos, CA 92069

Serial Number Book: Reference Book for Metalworking Machinery, Hearst Business Media Corporation, 29516 Southfield Road, Southfield, MI 48086

Surplus Record—Index of Available Capital Equipment, 20 North Wacker Drive, Chicago, IL 60606

The Last Bid, The Last Bid International Equipment Exchange, 6689 Peachtree Industrial Boulevard, Suite P, Norcross, GA 30092

The Truck Blue Book, National Market Reports, Inc., 300 West Adams Street, Chicago, IL 60606

Thomas' Register of American Manufacturers and Thomas' Register Catalog File, Thomas Publishing Company, One Penn Plaza, New York, NY 10001

Used Equipment Directory, 70 Sip Avenue, Jersey City, NJ 07306

Metalworking Machinery Manufacturers (Partial List)

Automatics

Brown & Sharpe Manufacturing Co.	Conomatic (Cone-Blanchard Machine Co.)	Warner & Swasey Co. New Britain Machine
National Acme		

Boring Mills

The Bullard Co.	Giddings & Lewis Machine Tool Co.
Wotan Machine Tools	

Chucking Lathes

Hardinge Brothers, Inc.	Warner & Swasey Co.

Drills

American Tool	Barnes Drill Co.
Buffalo Forge Co.	Burgmaster Div., Houdaille Industries

Forklift Trucks

Baker Material Handling Co.	Towmotor Corp.	Raymond Engineering, Inc.
Crown Industrial Products	Clark Equipment Co.	
Hyster Co.	Datsun	Toyota Machinery

Grinders

Brown & Sharpe Manufacturing Co.	Okomoto Corp.	Bryant Grinder Corp. (Van Norman)
Clausing Industrial, Inc.	Cone Blanchard Machine Co.	
Gorton Machine Co.	K. O. Lee Co.	Cincinnati Milacron
Landis Tool	Mitsui Seiki, Inc.	

Lathes

American Tool, Inc.
The Bullard Co.
Clausing Industrial, Inc.
LeBlond Machine Tool Co.
Lodge & Shipley Co.
Okuma Machinery, Inc.
Warner & Swasey Co.

Bardons & Oliver, Inc.
Cincinnati Milacron
Jones & Lamson Machine Co.
Logan Lathes,
Powermatic Div.

Mori Seiki U.S.A., Inc.
Pratt & Whitney
Gisholt Machine Co.

Machining Centers

Brown & Sharpe
LeBlond Machine Tool Co.
Okuma Machinery, Inc.

Warner & Swasey Co.
Burgmaster Div.,
Houdaille Industries
Mori Seiki

Mitsui Seiki
Pratt & Whitney
Giddings & Lewis
Cincinnati Milacron

Milling Machines

Bridgeport Machines
Kearney & Trecker Corp.
Tree Machine Tool Co., Inc.

Cincinnati Milacron
Okuma Machinery, Inc.

Presses

E. W. Bliss Co.
Dake Presses
Niagara Machine & Tool Works

Dries & Krumpf Mfg. Co.
Verson Allsteel Press Co.
Clearing Div., U.S. Industries

Danly Machine Corp.
Johnson Press Inc.
V & O Press Co., Inc.
Cincinnati Milacron

Press Brakes

Dries & Krumpf Mfg. Co.
Pacific Press & Shear Co.
Verson Allsteel Press Co.

Cincinnati Incorporated
The Lodge & Shipley Co.

Punch Presses

Alva Allen Industries
Johnson Press Inc.

Benchmaster Products Inc.
Niagara Machine & Tool Works

Saws

Armstrong-Blum Mfg. Co
W. F. Wells & Sons, Inc.

Hem Incorporated

Shears

Buffalo Forge Co.
Niagara Machine & Tool Works

Pacific Press & Shear Co.
Cincinnati Incorporated

Wysong & Miles Company
Peck, Stow & Wilcox

Welders

Linde Tool & Engineering Co.
Airco Welding Products

Sciaky Bros., Inc.
Miller Welding & Machine Co.

Lincoln Electric Co.

In the preceding pages, we have listed names of manufacturers and dealers who manufacture and/or sell machinery. It is not our intent that this be an endorsement or recommendation of their product to the exclusion of others. There are many fine companies whose names are not listed here who produce equal, if not superior, equipment. The names of the companies used were selected at random.

If you should encounter foreign-made machinery where a dealer cannot be found, the best avenue of approach is to check with the client's maintenance department. (They have to buy parts from someone.) If all else fails, you should contact the consulate of the country where the machine was manufactured.

To enhance your appraisal career, we have included a list of recommended readings. Many are available at public libraries or at a reasonable cost from the publisher.

In conclusion, your pricing and reference material library for machinery and equipment is one of the greatest tools for establishing and supporting an opinion of value. Utmost care should be taken to ensure that materials are kept up to date and filed in an orderly, systematic manner.

Recommended Readings

Bonbright, James C., *Valuation of Property*, vols. 1, 2, and 4, reprint by Michie Company, Charlottesville, Virginia, 1965.
Cottle, S., R. F. Murray, F. E. Block, *Graham and Dodd's Security Analysis*, 5th ed., McGraw-Hill, New York, 1988.
Daniells, Lorna M., *Business Information Sources*, University of California Press, Berkeley and Los Angeles, 1985.
Dell'isola, Alphonse J., and Stephen J. Kirk, *Life Cycle Cost Data*, McGraw-Hill, New York, 1983.
DeMarco, T., *Controlling Software Projects*, Yourdon Press, New York, 1982.
Graham, Frank, and C. Buffington, *Power Plant Engineers Guide*, 3rd ed., Audel, New York, 1983.
Grout, E. L., W. G. Iresou, and R. S. Leavenworth, *Principles of Engineering Economy*, Wiley, New York, 1982.
Houghton-Alico, D., *Creating Computer Software User Guides*, McGraw-Hill, New York, 1985.
Kennedy, Raoul D., *California Expert Witness Guide*, California Continuing Education of the Bar, Berkeley.
Marston, A., R. Winfrey, and J. C. Hempstead, *Engineering Valuation and Depreciation*, The Iowa University Press, Ames, Iowa, 1953.
McCarthy, G. D., and R. E. Healy, *Valuing a Company*, Ronald Press, New York, 1971.
M & E Appraiser, News Letter of the National Machinery and Equipment Committee, American Society of Appraisers, Washington, D.C., 1987.
Parker's Evidence Code of California, Parker & Son Publications, Los Angeles.
Petroleum Extension Service, *A Primer of Offshore Operations*, The University of Texas at Austin, Austin, Texas.
Petroleum Extension Service, *A Primer of Oilwell Drilling*, The University of Texas at Austin and International Association of Drilling Contractors, Austin, Texas.
Petroleum Extension Service, *A Primer of Oilwell Service and Workover*, The University of Texas at Austin and Association of Oilwell Servicing Contractors, Dallas, Texas.

Petroleum Extension Service, *A Primer of Pipeline Construction,* The University of Texas at Austin and Pipeline Contractors' Association, Dallas, Texas.

Smith, Gerald W., *Engineering Economy,* 3rd. ed., Iowa State Press, Ames, Iowa, 1979.

Standard Industrial Classifications Manual, Office of Statistical Standards, Superintendent of Documents, United States Government Printing Office, Washington, D.C.

Trott, A. R., *Refrigeration and Air Conditioning,* McGraw-Hill, New York, 1981.

7

Depreciation Theory

John Alico, P.E., FASA
*President, Alico Engineers & Appraisers
Birmingham, Michigan*

The purpose of this chapter is to present an approach to the determination of depreciation as it applies specifically to the appraisal of machinery and equipment.

Let us define a machine. What is a machine? Perhaps the most direct and simplistic definition is that a *machine* is a device which is acquired to perform a specific predetermined function. It is used singly or in combination with other items of machinery and equipment to enhance the productivity of an operating facility.

Estimating the amount of depreciation, expressed in dollars, is an integral function in the appraisal process when determining the value for machinery and equipment. *Value* is defined as the present worth of future benefits arising out of ownership to typical users or investors.[1] The practical appraiser is concerned with four general classifications of property: real, personal, tangible, and intangible. Real estate is considered as real, tangible property, while machinery and equipment are classified as personal, tangible property. There are many available references that describe the procedures for the appraisal of real estate.

---

[1] Bryl N. Boyce, *Real Estate Appraisal Terminology*, Balling Publishing, Cambridge, Massachusetts, 1975, p. 215, sponsored jointly by the American Institute of Real Estate Appraisers and The Society of Real Estate Appraisers.

There is, however, a relative dearth of information on the subject of appraising machinery and equipment.

An item of machinery or equipment is usually purchased in new or used condition to satisfy a utilization need. The investment for the purchase of the machine is justifiable on the premise that the part or product it will produce is a necessary element to the overall profitability and production operation of the purchaser. It must be assumed that the purchaser is a knowledgeable operator who is aware of the current levels of technology and the state of the art which prevail in a particular industry. In addition, the purchaser should be thoroughly familiar with the various kinds of machinery and equipment which are logical to acquire and use.

It is also necessary to know and understand the difference between price, cost, and value. Price and cost refer to an amount of money asked or actually paid for a property, and this may be more or less than its value.[2] *Value*, then, is what a property is worth. *Cost* is the price paid for the item of property plus freight, taxes, installation, and other expenses incurred by the purchaser in the acquisition of the property.[3] The cost of a property is not necessarily equal to its value. On the other hand, cost is considered as an evidence of value, and in establishing the value of property, it is customary to investigate both original cost and replacement cost.

In deriving an opinion of appraised value it is acceptable procedure to obtain the replacement cost new. For accounting purposes, however, original cost to the present owner provides the basis for book value considerations. It is readily evident that there is a difference in the value derived as an appraisal opinion and a value required for accounting purposes. It should be noted that there is a difference between historical cost and original cost.

Historical cost is the actual or first cost of a property at the time it was originally constructed and placed in service. It should not be confused with *original cost*, the latter term more properly being used to designate the actual cost to the present owner, who may have purchased the property at a price more or less than the historical or first cost. In an assembled property, such as a public utility, the historical cost as of any date means the first cost as defined, plus all subsequent additions and

---

[2]Henry A. Babcock, *Appraisal Principles and Procedures*, Richard D. Irwin, Homewood, Illinois, 1968, p. 95.

[3]Thomas L. Roberts, "Machinery and Equipment Appraisal in Cases of Eminent Domain," *The Appraisal of Machinery and Equipment*, American Society of Appraisers, Monograph no. 2, Washington, D.C., 1969, p. 83.

betterments, less deductions.[4] In summary, the historical cost is the retrospective estimate of the reproduction cost of a duplicate property.[5]

In appraisal practice, the pricing and costing of an item of property on any basis (original cost, replacement cost, or reproduction cost) is one of the first steps in the valuation process. This is recognized as the upper limit of value and includes the cost of installation plus any applicable indirect cost, e.g., engineering and "run-in" costs.

In practical appraising, the basic premise is to determine the value, expressed in dollars, of a property. There are many kinds of value which apply to machinery and equipment. These include value-in-place, insurable value, fair market value, salvage value, liquidation value, and scrap value. An appraisal is an estimate of value arrived at by an analysis of data developed by the appraiser.[6]

Accepted appraisal procedures have been taught by recognized real estate appraisal societies, which formally educate, test, and accredit appraisers. The procedures and accreditations have validity because they are recognized by the courts. Specialized knowledge and abilities are required of the appraiser, which are not possessed by the layperson.

The American Society of Appraisers recognizes and is concerned with all classes of property, including machinery and equipment.[7] The appraisal of this kind of property follows a logical process of fact-finding and analysis. The American Society of Appraisers has given creditability to the discipline of machinery and equipment appraising by sponsoring formal courses, publishing papers, and making educational cassettes available in its Audio-Library program.

Appraisal Procedure for Machinery and Equipment

The first step in appraising machinery and equipment is to make a thorough inspection of the property. This is required for purposes of identification and noting the physical condition of the property. Data and information for the determination of depreciation considerations are

---

[4]Boyce, *Real Estate Appraisal Terminology*, p. 215.

[5]Babcock, *Appraisal Principles and Procedures*, p. 95.

[6]William N. Kinnard, Jr., *Industrial Real Estate*, Society of Industrial Realtors, Washington, D.C., 1967, p. 392.

[7]American Society of Appraisers, *The Principles of Appraisal Practice and Code of Ethics*, Washington, D.C., March, 1977, p. 3.

obtained as part of the inspection. An item of machinery and equipment is usually bought for a specific purpose.

Lathes, milling machines, grinders, drills, and planers are used for metal removal. Mechanical and hydraulic presses, rolling mills, and draw benches are used for metal forming. Band saws, cutoff saws, and power shears are used for metal cutting. In addition to these standard items, machinery and equipment appraisers encounter specially designed installations such as pressure vessels, various types of air compressors, chemical process towers, mining equipment, and a variety of materials handling and transportation vehicles.

Exotic, but extremely practical items, such as robots, laser beams, computers, and computerized, numerically controlled devices with digital readouts, are appearing more frequently and, in fact, have become fairly common. Specific industries such as plastics, cement production, and woodworking have their own individual complement of machinery and equipment. The fact that these items of machinery and equipment are special-purpose and employed at their designed use indicates the application of the cost method in the valuation process.

The basic steps in the *cost method* to suit machinery and equipment appraisal practice are as follows:

1. Obtain the cost to replace (or reproduce) the basic machine new. Replacement cost of a property is the estimated cost of replacing the service of the existing property by another property, of any type, to achieve the most economical and preferred service, but at prices as of the date specified. This is, generally, the effective date of the appraisal. Reproduction cost of a property is the estimated cost of reproducing substantially the identical property at a price level as of the date specified.

2. Determine the physical depreciation due to wear and tear and exposure to the elements. Note the condition of the item under appraisement. Has it been well maintained? What are the noticeable effects of usage on the item? Is wear and tear evident as worn or broken parts? Is the machine loose and noisy? Are these physical deficiencies curable, and can the property be restored to its undamaged condition? If the appraiser can determine the cost-to-cure, this could be a measure of physical depreciation caused by utilization or exposure to the elements. If the condition of the machine is incurable, it is no longer useful and it has scrap value only.

3. Determine the remaining useful life of the property. Every new machine has a life expectancy, which is called its *economic life*. The ap-

praiser must estimate the effective age of the machine based on observed condition.

A suggested reference table, such as Table 7.1, may be utilized as a general basis for relating depreciation, condition, and remaining useful

Table 7.1. Depreciation Reference Table

Depreciation, %	Condition	Remaining useful life, %
	New	
0	New, installed, and unused property in excellent condition	100
5		95
	Very Good	
10	Like new, only slightly used, and not requiring any replacement of parts or repairs	90
15		85
	Good	
20	Used property, but repaired or renovated and in excellent condition	80
25		75
30		70
35		65
	Fair	
40	Used property which requires some repairs or replacement of parts such as bearings	60
45		55
50		50
55		45
60		40
	Usable	
65	Used property in operable condition, but considerable repairs or replacements such as motors or elements required	35
70		30
75		25
80		20
	Poor	
85	Used property requiring major repairs such as replacement of moving parts or main structural members	15
90		10
	Not Salable or Scrap	
97.5	No reasonable prospect of being sold except for the recovery value of its basic material content	2.5
100		0

life. The appraiser cannot use the table as a panacea for resolving every machinery and equipment depreciation analysis because the physical condition of every machine is not the same, and the table does not include consideration for obsolescence of any kind.

Depreciation—General

Depreciation may be interpreted to mean a cost of operation or physical condition. Generally, it is defined as a loss from the upper limit of value, which is the *replacement cost new, installed*. Case law reveals such definitions as "the decline in the value of an asset due to such causes as wear and tear, action of the elements, obsolescence and inadequacy"; "deterioration arising from age and use of improvements due to better methods, more economical and efficient designs, innovations and general advance in the art, notwithstanding reasonable current repairs"; and "physical and functional (depreciation) including wear and tear of property by earth's processes, supersession and obsolescence of machines and structures by progress." A concise statement in this category: "depreciation, broadly speaking, is the loss, not restored by current maintenance, due to all factors causing ultimate retirement of property, such as wear and tear, decay, inadequacy, and obsolescence."

Physical condition alone is neither depreciation nor the sole measure of depreciation, in the sense of either cost or value.[8] Machinery and equipment also undergo a loss from the upper limit of value as a result of age, inadequacy, obsolescence, and use. This premise is based on a constant price level, and the degree of loss is subject to adjustments due to price fluctuations. Age is a contributing factor to depreciation of a machine because a portion of its useful life has been consumed. This is reflected in a reduction of the present worth of the future benefits to be derived from ownership of the property.

It is generally reasonable to conclude that the value of an item of machinery and equipment will be at its highest when it is new, unused, installed, and ready for use.

Useful Life Expectancy

There was a time in our economic history when the useful, or service, life expectancy of a machine was quoted by manufacturers as 16⅔

---

[8]Anson Marston, Robley Winfrey, and Jean C. Hempstead, *Engineering Valuation and Depreciation*, Iowa State University Press, Ames, Iowa, 1968, p. 181.

years. This applied to any machine: radial arm drills, lathes of all kinds, planers, milling machines, boring mills, and saws, to name a few. Even a superficial examination will show that it is highly improbable for these machines to depreciate at the same rate when their utilization and application vary so widely. Inasmuch as the interest rate during that period was about 6 percent and there was no inflation, it would be reasonable to conclude that the depreciation rate was comparable to the annual reserve set aside to provide for the replacement of the asset when the machine was retired.

Actually, the useful life expectancy of a single machine varies widely. By definition, the *service life* of a property (machine) is that period of time (or service) extending from the date of its installation to the date of its retirement from service.[9] However, many things may affect the useful life of property. They include:

How often it is used

How old it was when acquired

How often it was repaired or renewed or its parts replaced

The climate in which it is used

The useful life can also be affected by technological improvements, progress in the arts, reasonable foreseeable economic changes, shifting of business centers, prohibitory laws, and other causes. All these should be considered before a useful life expectancy for the machine is determined.[10]

The useful life for the same type of machine varies from user to user. When the useful life of a machine is determined, it should be based on operating conditions which exist in the plant where it is used.

The guidelines set forth by the Internal Revenue Service in their Publication 534 (revised November 1983) "Depreciation" emphasize the premise that well maintained items of machinery and equipment have relatively low rates of depreciation. The most important point brought out by the IRS is that like items of machinery and equipment do not necessarily depreciate at the same rate. This conclusion can be substantiated by a comparison of the offerings and listings of used machinery made by dealers or advertised in trade journals.

The importance in the consideration of service age as related to depreciation is that a machine, when originally put in service, has 100 per-

---

[9]Ibid., p. 7.

[10]U.S. Department of the Treasury, *Depreciation*, Internal Revenue Service Publication 534 (revised), November 1983.

cent of its useful life remaining at that time and the present worth of its probable future service is at its highest. As the service age increases, there is a corresponding decrease in the value of the property because the service life remaining is shorter and the present worth of its probable future service is reduced accordingly.

Physical property decreases in value with age and use because it thereby suffers a decrease in the amount and character of the future service it can render before it will be retired.[11] This was the basic premise for organizing Table 7.1.

Effects of Condition on Depreciation

Age cannot be considered as the sole basis for depreciation. *Deterioration* is defined as impairment of condition. Deterioration is one of the causes of depreciation and reflects the loss in value brought about by wear and tear, disintegration, use in service, and the action of the elements.[12] It is evident that condition is a factor to be considered in determining depreciation for appraisal purposes.

Example 1

Three Machines, Three Conditions

Consider a situation where three new identical machines are delivered on the same day to three different manufacturers in the same geographic area. The first manufacturer (Case 1) subscribes to the philosophy that any machine the company buys must produce and that downtime is an intolerable loss of revenue. Maintenance in this case is minimal, and the effects of usage (wear and tear) become apparent relatively early in the service life of the machine. At the end of a three-year period of intensive use, the remaining useful life expectancy of this machine can be directly related to its physical condition and its incapability to produce in accordance with design specifications. Obviously, hard usage has shortened its effective useful life, and its physical depreciation rate has been relatively high.

The second machine (Case 2) was delivered to an operator who subscribes to the philosophy of regular periodic maintenance checks. Part of this program calls for a general overhaul after specific periods of usage, say, every 500 service hours. Worn parts such as gears and bearings are replaced, and in

---

[11]Marsten et al., *Engineering Valuation and Depreciation*, p. 7.

[12]*Appraisal Terminology and Handbook,* 4th ed., American Institute of Real Estate Appraisers, Chicago, 1975, p. 53.

effect the machine is rebuilt to a like-new condition. After three years of service, it would be reasonable to assume that the condition of this unit is better than that of Case 1. The age is the same, but because of its improved condition, the remaining useful life expectancy will undoubtedly be longer. The physical depreciation rate would, therefore, be lower than that of the machine in Case 1.

The third machine (Case 3) was delivered to a manufacturer who did not require its utilization at the time and decided to store it. Circumstances were such that the item remained in storage for a three-year period. This unit could be described as being in a new-excellent condition and having a 100 percent useful life expectancy despite the fact that the serial number indicates it is three years old.

It is obvious from Example 1 that condition is a factor to be considered in the determination of depreciation. Moreover, condition has a direct effect on remaining service life. In order to determine the relationship between condition, depreciation, and remaining useful life, it is necessary to define the various forms of condition. Two or more appraisers could inspect an item of machinery and equipment, and it is quite probable that each will arrive at a different description of condition. It is necessary, therefore, to establish some definitions of condition for the purpose of attempting to reach basically similar concepts and thus eliminate the subjective factors in the determination of depreciation.

Practical appraisers often base depreciation on their individual interpretation of the physical condition of the property with the result that it is fairly common practice for appraisers to use physical condition as a direct index of value of the property. Physical condition, alone, is neither depreciation nor the sole measure of depreciation, in the sense of either cost or value.[13] While this is true, it must be emphasized that physical condition can be defined and measured as to its contribution to depreciation.

Definitions of Condition

Condition, however, is a characteristic that can be determined only through observation. The appraiser should have a clearly established understanding with the client as to the various definitions of condition. The subject of condition can be an area for disagreement. Several individuals could inspect an item of equipment and have differing descriptions as to its condition. It would be good to consider making definitions

---

[13]Marsten et al., *Engineering Valuation and Depreciation*, p. 7.

of condition a part of every contract for the appraisal of machinery and equipment. The definitions of condition should also be included in the appraisal report. A suggested set of definitions is given below.

Very Good (VG)

This term describes an item of equipment in excellent condition capable of being used to its fully specified utilization for its designed purpose without being modified and without requiring any repairs or abnormal maintenance at the time of inspection or within the foreseeable future.

Good (G)

This term describes those items of equipment which have been modified or repaired and are being used at or near their fully specified utilization.

Fair (F)

This term describes those items of equipment which are being used at some point below their fully specified utilization because of the effects of age and/or application and which require general repairs and some replacement of minor elements in the foreseeable future to raise their level of utilization to or near their original specifications.

Poor (P)

This term is used to describe those items of equipment which can be used only at some point well below their fully specified utilization, and it is not possible to realize full capability in their current condition without extensive repairs and/or the replacement of major elements in the near future.

Scrap (X)

This term is used to describe those items of equipment which are no longer serviceable and which cannot be utilized to any practical degree regardless of the extent of the repairs or modifications to which they may be subjected. This condition applies to items of equipment which have been used for 100 percent of their useful life or which are 100 percent technologically or functionally obsolete.

Obsolescence in Machinery and Equipment

It should be emphasized that in addition to age and condition, other factors, particularly obsolescence, should also be considered in any attempt to appraise a machine and apply depreciation considerations. Obviously, the consideration of depreciation based only on observed condition cannot be a practical basis for estimating the value of a machine.

Many appraisers have used observed physical condition as a direct index of value in machinery and equipment valuation. However, this is not the sole basis for depreciation considerations. Along with age and condition, obsolescence and utility are also contributing factors and should be part of the total depreciation analysis.

There are three kinds of obsolescence which may affect the value of a machine: technological, functional, and economic. In real estate, functional and economic obsolescence are defined and considered in appraising the value of property. Obsolescence is generally recognized as contributing to the loss from the upper limit of value of a property. In machinery and equipment, where technology and the state of the art are continually changing in design, materials of construction, and manufacture, it is necessary to consider the degree of obsolescence which these changes contribute to the loss of value.

Technological Obsolescence

Technological obsolescence relates to the difference between the design and materials of construction used in present-day machines compared with the machine under appraisement. Size and overall floor space requirements are other examples of technological obsolescence. For example, a replacement machine has a floor space requirement of 50 ft^2 compared with 60 ft^2 for the older machine being evaluated. This is a difference of 10/60 or 16.6 percent. Management works toward increasing productivity per square foot of plant area, and floor space requirements of machines are important considerations which relate to the desirability of owning a particular machine.

Functional Obsolescence

Functional obsolescence has to do with the difference in production rates and other capability characteristics between a new machine and the machine being evaluated. Direct labor requirements would be another consideration in this respect. A new machine might produce 45 units per time frame compared with 40 units produced in the same period by the machine being appraised. This is a difference of 5/40 or 12.5 percent.

Items of machinery and equipment are usually designed for or adapted to a specific use. This can be defined as the highest and best use for the subject item and the most profitable likely use to which a property can be put. The opinion of such use may be based on the highest and most profitable continuous use to which the property is adapted and needed or likely to be in demand in the reasonably near future. The ability of a property to be utilized at its highest and best use would have some relationship to value. Any utilization less than highest and best use would be a contributory factor to depreciation because this represents a loss from the upper limit of value. This limitation in use could be described as functional obsolescence.

Economic Obsolescence

Economic obsolescence deals with influences external to the machine itself. It is defined as the impairment of desirability or useful life arising from economic forces, such as changes in optimum use, legislative enactments which restrict or impair property rights, and changes in supply-demand relationships. These factors, too, can be measured and expressed in percentages of the subject machine's productivity or potential use. It is possible for a machine to be 100 percent economically obsolescent.

A case in point would be the finishing line for 12-cylinder automotive engines for passenger automobiles such as those used in the original Lincoln Zephyr automobile. Currently, the government requirement for improved mileage performance in American-made automobiles is reducing the output of eight-cylinder engines, and the production lines for these products are rapidly approaching 100 percent obsolescence. When this occurs, the remaining useful life of these machines is zero regardless of age or condition.

However, for purposes of illustration, assume that a valued customer has moved out of profitable shipping range or gone out of business and the competitive situation is such that a replacement account cannot be found. Also assume that the equipment used for the former customer now has only a 90 percent effective production schedule. This would be measured as having a 10 percent economic obsolescence. All the obsolescence considerations are additive; hence, if technological obsolescence is 16.6 percent, functional obsolescence is 12.5 percent, and economic obsolescence is 10 percent, the total is

39.1 percent, which must be added to physical depreciation in order to determine the total loss of value.

This, then, is the amount which has to be charged against the machine which is being appraised. Assume, further, that the subject unit has a replacement cost new, installed, of $145,000, and that depreciation due to age and the effects of wear and tear amounts to 35 percent. Obsolescence is considered as follows:

Replacement cost new, installed		$145,000
Depreciation due to age and use	35.0%	
Obsolescence considerations	39.1%	
Minus total percent loss of value	74.1%	or 107,445
Value in place		$ 37,555

The procedure just described relates to valuation appraisals. The purpose of outlining the method is to give the appraiser as much factual data as possible to substantiate conclusions of value. The evaluation of machinery and equipment is highly complicated because there are many variables and, as value is a judgment quantity, the appraiser should be familiar with methods which could reduce possible error in the results. In any case, the values determined by one appraiser are not likely to agree, exactly, with those stated by another, particularly when an item of machinery is the subject.

Discussion of the Market Data Approach as Applied to Depreciation

In appraising machinery and equipment, an appraiser may go to the marketplace, such as a dealer in used machinery or offerings listed in used machinery directories,[14,15] and obtain what pricing data prevails as the market price at the time of inquiry. The appraiser then subtracts this offering price from the replacement cost new, installed, of the property. Then the appraiser labels the result as value-in-place or fair market value. The difference between the replacement cost, installed,

---

[14]*Industrial Machinery News,* monthly publication of Hearst Business Media Corp., 29516 Southfield Road, Michigan 48037.

[15]*Locator of Used Machinery and Equipment,* published monthly by Machinery Dealers National Information Systems, Inc., 1110 Spring Street, Silver Springs, Maryland 20910.

and the offering price is called *depreciation*. The depreciation in this case has no bearing on the estimated useful life remaining, the condition, or the utility of the item being appraised, and, therefore, such practice may be considered as improper.

The machine being offered for sale at a price set by the dealer may often be purchased at a lower price arrived at by negotiation or under auction sale conditions. Conversely, a manufacturer may require a machine to overcome a production crisis. In this kind of situation, a machine may be sold for a price higher than its appraised value. Similar situations occur when a particular new machine requires an unusually long lead time for delivery. Metalworking operators will often pay a premium for a similar machine in the used condition in order to obtain immediate delivery.

The prices in either of the two cases described above may not have a direct relationship to value, but because the machines were sold, market data was thus generated. Machines offered for sale are rarely of the same age, condition, and utility as the machine under appraisement. Consider the impracticability of relating a machine installed in a manufacturing facility in Detroit, Michigan, to a machine being offered for sale by a dealer in Los Angeles, California. The obvious question would be: How is it possible to substantiate the premise that the two machines are equal in condition and utility?

Reference to trade magazines will show the variations in offerings of basically similar machines by dealers in different locations. This is not a criticism of the used machinery dealer. The used machinery industry is a time-honored and wholly legitimate business which fulfills an economic need. Variations in offerings could be the result of differences in condition, supply-and-demand situations, or regional economic variables. It may be argued that salability could be related to condition, and that the offering price, if firm, may reflect the amount of depreciation in a machine, but this would have to be proven. It is the appraiser's responsibility to include this proof in the report.

It should be noted that offerings are, almost always, higher than the cost of the machine to the used market dealer. Costs may include transportation to the dealer's shop or storeroom, interest on money, advertising costs, and administrative and overhead expenses. These must be recovered. When the machine is sold, there must be some profit in the transaction if the dealer is to stay in business. None of these items directly contribute to the depreciation of a machine. Not all used machines are sold in the used equipment market. Negotiated sales of machines between an owner and a purchaser-user are rarely recorded, but they do occur. If the market is to be considered as a basis for depreciation, it follows that all sales should be included, auctions, open market, and private.

Accounting Methods Used to Determine Depreciation

It is also fairly common practice for appraisers to base valuations on depreciation methods which actually apply to original cost allocations. There are various methods and procedures which are used for cost accounting purposes, and these may be divided into two groups. The first group is considered to be a noninterest procedure for estimating depreciation of single units of property. This group includes *straight-line, declining-balance,* and *sum-of-the-years'-digits* depreciation methods.

The second group consists of procedures which are based on interest theories. These include the *sinking-fund depreciation method* and the *present-worth theory.* All the methods identified in the two groups are based on formulas and considerations which will produce a result for any given period of time regardless of age, condition, utility, and the possible occurrence of obsolescence. The results are not actual appraisals and are generally called *book values.*

Noninterest Procedures

Straight Line. In straight-line depreciation, a fixed rate is applied each year (or given period of time) usually against the original cost. This procedure is formulated below.

$$D = R \times C$$

where D = depreciation, dollars
R = fixed depreciation rate × number of time periods
C = cost (historical or otherwise)

Assume an original cost of $40,000 and a 20-year life. Depreciation will be 5 percent per year so that in seven years, the total depreciation will be 35 percent of $40,000, or $14,000.

Declining Balance. The declining-balance depreciation method is a procedure for depreciating an asset by use of a fixed percentage applied to the successive balances remaining after previously computed amounts of depreciation have been deducted. In other words, the annual depreciation charges are determined on the basis of some fixed multiple (125 percent, 150 percent, or 200 percent) of the straight-line depreciation rate as applied to the undepreciated balance of the property.[16]

For the 125 percent declining-balance method, the appropriate annual rate to apply would be 1¼ times the straight line rate, or 6.25 per cent; for the 150 percent declining-balance method, it would be 1½ times the straight-line rate, or 7½ percent; and for the 200 percent declining-balance method, the proper annual rate to apply would be 10 percent. The declining-balance schedule showing the annual depreciation for a $40,000 machine over 20 years is presented in Table 7.2.

Sum of the Years' Digits. In the sum-of-the-years'-digits method, the annual depreciation charges are computed on the basis of a fraction which changes over time (declines) applied to the initial depreciable cost of an asset. The numerator of the fraction represents the remaining

Table 7.2. Declining Balance Schedule at One Hundred and Twenty-Five Percent

Year	Balance	Annual depreciation	Cumulative depreciation
0	$40,000	—	—
1	37,500	$2,500	$ 2,500
2	35,156	2,344	4,844
3	32,959	2,197	7,041
4	30,899	2,059	9,100
5	28,967	1,931	11,031
6	27,156	1,810	12,841
7	25,459	1,697	14,538
8	23,868	1,591	16,129
9	22,376	1,492	17,621
10	20,977	1,398	19,019
11	19,666	1,311	20,330
12	18,437	1,299	21,629
13	17,285	1,152	22,781
14	16,204	1,080	23,861
15	15,191	1,013	24,874
16	14,242	949	25,823
17	13,351	890	26,713
18	12,517	834	27,547
19	11,735	782	28,329
20	11,001	733	29,062

[16]Boyce, *Real Estate Appraisal Terminology*, p. 215.

Depreciation Theory

useful life of the asset and changes each year, while the denominator represents the sum of all the years' digits and remains constant. The generalized formula for this fraction is[17]

$$\frac{N}{N(N+1)/2}$$

An example of how the sum-of-the-years'-digits method of depreciation applies is shown in Table 7.3 for a machine acquired at a cost of $40,000 and depreciated over a 20-year period of time.

The methods described above do not involve the theory of interest and have their application primarily in cost allocations as well as approximations in engineering economy studies.

Neither the declining balance nor the sum-of-the-years'-digits method have enough merit to warrant their use in practical appraisal procedures.

Table 7.3. Sum-of-the-Years'-Digits Schedule

Year	Annual depreciation	Cumulative depreciation	Fraction
1	$3,809	$ 3,809	20/210
2	3,619	7,428	19/210
3	3,429	10,857	18/210
4	3,238	14,095	17/210
5	3,048	17,143	16/210
6	2,857	20,000	15/210
7	2,667	22,667	14/210
8	2,476	25,143	13/210
9	2,286	27,429	12/210
10	2,095	29,524	11/210
11	1,905	31,429	10/210
12	1,714	33,143	9/210
13	1,524	34,667	8/210
14	1,333	36,000	7/210
15	1,143	37,143	6/210
16	952	38,095	5/210
17	762	38,857	4/210
18	571	39,428	3/210
19	381	39,809	2/210
20	190	40,000	1/210

[17]Ibid., p. 215.

Two methods which employ the interest theory are the sinking-fund and present-worth theories as applied to single units. Depreciation sinking funds are rarely used in industry[18] and, for a while, were in general use in the public utility field. The method is still used extensively in engineering economy studies as a basis for selection of alternative proposals.

Procedures Based on Interest Theories

Sinking Fund. In the sinking-fund depreciation method, the total allocation of the base to any date is equal to the corresponding accumulation (of annual payments to the sinking fund and compound interest thereon) in a hypothetical equal-annual-year-end-payment sinking fund, in which the total accumulation at the end of the service life of the unit would be just equal to the depreciable base of the unit. The sinking-fund method, although based upon a well established principle of compound interest, is not based upon a premise related to the consumption of usefulness of the property.[19] For this reason, it should not be considered as a basis for depreciation considerations in practical appraisal procedure.

Present Worth. The present-worth method of depreciation is similar to the sinking-fund method but is based on the fundamental concept of value. The principle is that the fundamental basis of value of a property is the present worth of its probable future services. In other words, the present-worth principle is that the value of a property, at any date during its service life, is the present worth at that date of the probable future operation returns yet to be earned through its probable future services.[20]

By definition, neither the sinking-fund nor the present-worth depreciation principles are applicable to the problem of determining an opinion as to the value of an item of machinery or equipment in the market or for collateral purposes.

---

[18]Kinnard, *Industrial Real Estate*, p. 392.

[19]Marston et al., *Engineering Valuation and Depreciation*, p. 7.

[20]Ibid., p. 7.

Summary and Conclusions

Evaluating business enterprises and industrial property today is an important appraisal function. It is essential for the appraiser to have some understanding of the engineering, economic, financial, legal, and management influences on value. Assume that a corporation purchases 10 acres of land at a cost of $10,000 per acre. This is an expenditure of $100,000 for 435,600 ft^2 on which a 100,000-ft^2 good class C building is erected at a total cost of $40 per square foot, or $4,000,000. Land improvements, such as parking lots, fencing, landscaping, and a rail spur cost an additional $350,000. The total cost of the real estate, in this case, amounts to $4,350,000. It is conceivable that this facility will house $7,500,000 worth of machinery and equipment, which could include office furniture and machines such as computers, word processors, copy machines, and electronic typewriters. Rolling stock, consisting of trucks and automobiles, could also be included.

The purpose of this chapter is to indicate the methods for determining the value of machinery and equipment as a separate asset in the overall worth of a business enterprise. To include machinery and equipment as a part of the real estate is fallacious because in the event of a dissolution of the enterprise or dispersal of the assets, the real estate is fixed while the machinery and equipment, for the most part, are removable. Furthermore, the real estate, including land, land improvements, and buildings and structures, would be sold as a unit, in place. Machinery and equipment, on the other hand, may be sold in units to many purchasers and removed for off-site use. The need to develop methods for evaluating machinery and equipment on a unit basis becomes apparent under the circumstances cited above.

Machinery and equipment are bought to fulfill a predetermined function or need. The value of a machine is in direct proportion to its utility. Machines are often special-purpose in design and, when employed at their highest and best use, approach their highest values. The equipment installed in the best managed and best maintained plants will depreciate in time due to wear and tear. Because machines have service lives which ordinarily last for relatively long periods, say 12 to 15 years, careful use and good maintenance may extend these useful lives for much longer periods.

Depreciation includes physical wear and tear, exposure to the elements, and technological, functional, and economic obsolescence. Changes brought on by technical improvements may cause a loss in value attributable to obsolescence, both technological and functional.

Economic obsolescence, due to loss of market, government edict, or other forces outside the property, is also a depreciation consideration because it results in the reduction or loss of the utilization of a property.

There is a distinct difference in the meaning of the word *depreciation* between the accountant's and appraiser's concepts of the term. It was shown in this chapter that in accounting practice depreciation is identified as the recovery of cost over the estimated useful life of the property. The primary objective is the determination of net income for tax reporting and other purposes. Therefore, depreciation is treated as an allocation of cost for the purpose of deducting expenses from current income. It is obvious, then, that accounting for depreciation is not a valuation procedure.[21] The practical appraiser is not concerned with accounting but is attempting to validate an opinion of the true value of the equipment. The appraiser's aim is to estimate the value of a specific machine at a particular time and to use depreciation as the difference between the replacement cost new and its value.

The measurement of economic life and depreciation is a most important but difficult step in the appraisal process. For equipment, it is particularly difficult because of the various types, grades, and uses of such property, and because market data cannot serve as a reliable basis for estimating depreciation. The relationship between age, condition, utilization, and remaining useful life demonstrated in this chapter should help the appraiser determine an opinion of value based on factual data.

---

[21]*The Appraisal of Equipment, Inventory, and Supplies,* Assessor's Handbook, AH571, Assessment Standards Division, Property Tax Department, California State Board of Equalization, April 1971.

8
Fair Market Value Concepts

Robert S. Svoboda, ASA
Senior Engagement Manager, American Appraisal Associates, Inc., Milwaukee, Wisconsin

Basic Concepts

This chapter addresses the valuation of machinery and equipment utilizing the three traditional approaches to value: cost, market, and income. The sections on the cost and market approaches discuss the valuation of individual items of machinery and equipment. The section on the income approach discusses conceptually the valuation of a business and illustrates that there may be additional economic obsolescence not measurable using the cost and market approaches when appraising machinery and equipment as part of a business.

All the concepts presented in this chapter have merit when appraising machinery and equipment, but there are some practical limitations which limit their universal use. Ideally it would be nice to utilize many of the ideas presented, but as most appraisers know, there is a limited amount of time in which to make an appraisal. In addition, many of the ideas depend on a sufficient amount of data to make an analysis, and many times the data are simply not available. Therefore, an appraiser must make some decisions as to how to approach the appraisal of machinery and equipment. The intent of this section is to present some ideas and examples as to how various concepts can be applied when time and data are available.

This chapter discusses value under the premise of continued use, which assumes the property will continue to be used for the purpose for which it was designed and built or to which it is currently adapted. The premise implies that the property will be retained at its present location for continued operation. It is generally appropriate under the following circumstances:

- The property fulfills an economic demand for the utility it provides.
- The property has remaining useful life.
- The continuation of the existing use is practical.
- An alternative use would not be feasible.

For machinery and equipment, this premise implies that the assets are installed, operating, and an integral part of the entity in which they are employed.

Generally the continued-use premise requires the appraiser to qualify the general definition of fair market value to further explain the assumptions and limiting conditions which affect the value conclusions. For example, appraisers are often asked to appraise machinery and equipment which operate as part of a business for continued use. According to Section 6.3, "Fractional Appraisals," of the American Society of Appraisers (ASA), *Principles of Appraisal Practice and Code of Ethics*, this situation would be considered a fractional appraisal; i.e., we are appraising a fraction separated from the whole property. This situation will probably result in a different value from what it would be if it is considered as an integrated part of the whole property. By using the continued-use premise, we are assuming that the property fulfills an economic benefit, which, for a business, is its earnings potential, and this needs to be qualified. In a fractional appraisal for continued use, the appraiser is assuming that the prospective earnings of the business are adequate to support the level of value concluded and a qualification to the general definition of fair market value is appropriate. In the case where earnings have been analyzed, the general definition of fair market value should be further qualified.

Throughout this chapter, we refer to the general definition of fair market value as defined elsewhere in this text, but it is qualified according to the premises used and the assumptions made. The Machinery and Equipment Committee of the American Society of Appraisers has adopted the term *fair market value-in-place* (in use) as being "the fair market value of an item including installation and the contribution of the item to the operating facility (and) presupposes the continued utilization of the item in conjunction with all other installed items." Basically this definition represents the value of an asset that is installed, operating, and contributing to the business in which it is employed. Implicit in this definition is the assumption that this value level of the individual assets can be economically justified. In other words, a prudent buyer can afford to purchase this asset at the indicated value. When appraising machinery and equipment fractionally (separate from the business), the fair market value of the asset is equal to the fair market value-in-place and the appraiser assumes a prudent buyer would pay this price. If the appraisal is being done within the context of the business, as a going concern, fair market value is often not equal to fair market value-in-place and other factors need to be considered, factors which would be considered by a prudent buyer. For the purposes of this chapter, fair market value-in-place is the value level derived from the cost and market approaches. It is the aggregate value of the individual items without regard to how the assets are used within the business in which they are employed. Later in the chapter, when we discuss the income approach to value a business, we see how to go from fair market value-in-place to fair market value.

The Cost Approach

The logic behind the cost approach is that a prudent investor would pay no more for a property than the cost of producing a substitute property with the same utility as the subject. Therefore, if a property is new, the current cost of producing that equivalent tends to establish the upper limit of value. When the property is not new, then the appraiser must deduct for the various elements of depreciation—physical deterioration and functional and economic obsolescence. This section discusses in detail the use of the cost approach applied to machinery and equipment.

Definition

There are numerous definitions of the cost approach, most of which are derived from the principle of substitution. The principle of substitution states "that a prudent purchaser would pay no more for ... property than the cost of acquiring an equally desirable substitute on the open market."[1] The principle applies to the cost approach when the pur-

---

[1] Bryl N. Boyce, *Real Estate Appraisal Terminology*, Ballinger, Cambridge Mass., 1975, pp. 201. Copyright 1975 by the American Institute of Real Estate Appraisers and The Society of Real Estate Appraisers. Reprinted with permission from Ballinger Publishing Company.

chaser can construct or buy a new substitute property with equivalent utility. The principle can be applied either to an individual asset or to an entire facility. For this discussion we will define the cost approach as follows:

> *Cost approach* is that approach which measures value by determining the current cost of an asset and deducting for the various elements of depreciation, physical deterioration and functional and economic obsolescence.

In its simplest form the cost approach is nothing more than current cost less all depreciation. When using the cost approach, the appraiser identifies the assets being appraised, develops the current cost of the subject, and subtracts all depreciation that makes the subject less desirable to own than if it were new. The factors that decrease value are expressed in percentages and converted to dollars or expressed directly in dollars. Thus the cost approach is summarized as follows:

Current Cost of Replacement or Reproduction New

Less	Physical deterioration
Less	Functional obsolescence
Less	Economic obsolescence
Results in	Fair market value-in-place

Determination of Current Cost

Development of current cost is discussed elsewhere in this book, but its use as part of the cost approach is discussed here. The first step in the cost approach is to determine the proper level of current cost, cost of reproduction, which is the cost of producing or constructing a property in like kind, or the cost of replacement, which is the cost of producing or constructing a property of equivalent utility. The normal cost elements included in either the replacement or reproduction cost estimates are all direct and indirect costs.

When two or more machines, or process units, are available with like utility and capacity, the one with the most favorable investment and production costs is regarded as the most desirable from an economic standpoint. Therefore, when estimating value by the cost approach, these factors must be carefully analyzed to properly establish the basis for developing the proper level of current cost.

In theory the development of reproduction cost is usually the first step in applying the cost approach, but only if it is the most practical and economical approach available to a prudent investor. If reproduction is not possible (perhaps specific construction materials are no longer available) nor technically feasible because of advances in the state of the art within an industry, the proper basis becomes the cost of replacement.

The *cost of replacement* is the proper cost basis for developing an opinion of value using the cost approach. The cost of replacement is the upper limit of value[2]—what the property would be worth to a prudent investor in the new and unused condition. Using the principle of substitution, a property can be worth no more than the cost (read replacement) of an equally desirable substitute in the new and unused condition.

When using the cost approach, the appraiser is comparing an existing facility to its modern counterpart. Improvements and changes in layout, design, materials, product flow, construction methods, and equipment size and mix make the modern equivalent more desirable from both a *capital* (cost to build) and *operating* (cost to operate) cost standpoint. Generally the cost of replacement for a modern substitute is lower than the cost of reproduction of the subject property. The difference represents one form of functional obsolescence—that due to excess capital costs. These excess costs represent the decreased capital investment required for a desirable substitute resulting from the improvements in technology and materials.

It is possible that replacement cost can exceed reproduction cost, especially when increased capital costs are offset by significantly reduced operating costs. This concept is discussed further in the section entitled "Functional Obsolescence," which follows.

The development of the replacement cost does not preclude development of the reproduction cost. Reproduction cost can be developed to estimate the amount of excess *capital cost* (reproduction cost minus replacement cost) or provide a basis to allocate or distribute various penalties. In some cases, an appraisal may be compared to others where original costs have been trended, so the development of reproduction cost will provide a basis of comparison.

Replacement cost is developed based on the principle of substitution, which takes the form of producing a substitute property with the same

---

[2] Do not confuse the upper limit of value with the upper limit of cost. Usually, reproduction cost is higher than replacement cost, so reproduction cost establishes the upper limit of cost. To move from cost to value put yourself in the position of buyer with a choice of paying reproduction cost or replacement cost for equivalent utility. All other things being equal, you will probably choose replacement cost because it represents an equally desirable substitute at a lower cost. Therefore, the value of assets in the new and unused condition is established by replacement cost.

utility and productive capacity. To apply the principle, the appraiser strives to obtain the current cost of that substitute. Whether establishing a replacement cost for an individual machine, an operating department, a process unit, or an entire facility, the cost estimate must conform to current design, engineering, technology, and materials of construction.

Example 1
Using Proper Level of Current Cost in Appraisal

Your company has been asked to appraise a 50,000 barrel per day petrochemical refinery in the Midwest, and your role is to appraise the steam generation facilities. You take a plantwide tour and ascertain that the steam generated is for process use only. You meet the boiler superintendent and later have a meeting to discuss the facilities. In your meeting you are provided with the following information:

Boiler number	Capacity, lb/hr	Chronological age, years	Type
1	15,000	46	Brick-set oil-fired
2	30,000	36	Field erected oil-fired
3	30,000	10	Package oil-fired
4	60,000	6	Package oil-fired

Boiler 1 is still capable of operating, but it has been considered excess since Boiler 4 began operation six years ago. Boiler 1 has not operated since.

Boiler Operations You discussed the operations with the superintendent and determined the following facts:
1. Boiler 2.
   a. Is a "swing" boiler.
   b. Runs occasionally during peak periods.
2. Boiler 4.
   a. Runs continuously.
   b. Maximized use.
3. Boiler 3.
   a. Operates in conjunction with boiler 4.
   b. Usually operates at or near capacity.
4. Peak periods are midsummer and midwinter.
   a. Peak demand summer: 100,000 lb/hr.
   b. Peak demand winter: 90,000 lb/hr.

The client has recently built a refinery of equal capacity in a neighboring state. The plant manager of that new facility will be on site later in the week, so you schedule a meeting with him to discuss a modern replacement plant

Fair Market Value Concepts

which you hope to use in your appraisal. During that meeting, you are provided with the following information:

Refinery is of equal capacity: 50,000 barrels per day.

Construction was completed three years ago.

The plant utilizes heat recovery on the process units to reduce plant-generated steam.

The new plant has 20,000 to 40,000 lb/hr package-type boilers, oil-fired.

Peak demands in summer and winter have virtually been eliminated because of the heat recovery.

Original cost of the boilers including all peripherals is $1,500,000.

The question then is "What is the proper level of current cost to be used in the cost approach?" The appraisal is being done for ad valorem tax purposes under the premise of continued use with the appraisal date being current. Assume all costs are midyear, and use 10 percent per year for trending.

You should begin by analyzing the available facts. Since the company just built a new refinery, you have the basis for establishing the replacement plant for the entire refinery and specifically the steam generation facilities. (Assume that the feedstock and the product mix are the same.) When you look at the boilers, you see a number of areas of obsolescence. Boiler 1 is brick-set oil-fired and represents excess construction. Although it is not mentioned specifically, it would be logical to assume that the use of heat recovery on the refinery process units represents an advance in technology resulting in reduced steam requirements. Your plant has a total steam capacity of 135,000 lb/hr, while the new plant has a total of 80,000 lb/hr, representing excess capacity. Since a prudent purchaser would pay no more for property than the cost of acquiring an equally desirable substitute, the proper level of current cost would be that cost for two 40,000-lb/hr package-type oil-fired boilers. In this example, that would be approximately $2,000,000 ($1,500,000 × 1.331 = $1,966,500 rounded to $2,000,000).

Although it will not be discussed, realize that the reproduction cost for four boilers of varying sizes and types is far greater than the $2,000,000 required for the modern steam generation facilities.

In this example, there is a minor problem that needs to be discussed. We have addressed the issue of plant-generated steam which has been reduced in a modern plant because of the use of heat recovery on the process units. Although there is a corresponding cost savings for the steam-generation facilities, there is probably a capital cost increase associated with the additional equipment needed on the process units to recover waste heat. The point here is that the replacement plant needs to be balanced from both a capital and operating cost standpoint. This is especially true when appraising integrated facilities such as oil refineries or other process plants.

Example 2
Showing Replacement Cost May Be Greater than Reproduction Cost

You have been asked to appraise Machine A with the following information:

Age = 6 years
No longer manufactured, replaced by model B
Last selling price, $100,000
Cost of reproduction new (determined by trending): $130,000
Capacity: 100 units per day

At first glance you might be tempted to use the reproduction cost as the starting point in the cost approach, but in your discussions you learn that the unit is no long manufactured and has been replaced by Machine B. You do some further investigation and find that Machine B is also rated at 100 units per day, but its cost is $160,000. The reason for the capital cost difference is an advance in technology that drastically reduces the energy required to produce each unit. On an annual basis, the subject requires 100,000 therms (1 therm = 100,000 Btu) of natural gas, whereas Model B requires only 75,000 therms, a difference of 25,000 therms. If we assume 50 cents per therm, Model B will save $12,500 per year, which will justify purchasing Machine B at the higher price.

In this example, the newer model is more desirable than the old one simply because it can produce an equivalent number of units per day with reduced energy consumption. The cost of a new machine is greater than the cost of the old machine, but it makes up the difference with reduced operating costs. Therefore the replacement cost is greater than the reproduction cost. Later, when we discuss functional obsolescence we show how to measure the additional depreciation attributable to the older model resulting from the reduced operating cost.

The Three Elements of Depreciation as Used in the Cost Approach

The three elements of depreciation recognized by virtually all appraisers are physical deterioration and functional and economic obsolescence.[3] They are discussed and illustrated as used in the cost approach.

Physical Deterioration. Once the proper level of current cost has been determined, deductions must be made for depreciation. Normally physical deterioration is considered first and is defined as follows:

> *Physical deterioration* is the loss in value resulting from wear and tear of an asset in operation and exposure to various elements.

Physical deterioration is simply wear and tear of an asset by its use. It is a result of past service experience and maintenance practice, exposure to the natural elements or the process area environment, internal de-

---

[3]Some machinery and equipment appraisers draw a distinction between technological obsolescence and functional obsolescence, while others consider them synonymous. Refer to Chapter 7 for further discussion.

fects from vibration and operating stress, and the effects of prolonged shutdowns, accidents, and disasters. Excessive deterioration often reduces tolerances on manufacturing equipment, resulting in higher product rejection and ultimately the inability to meet production standards. It may require continual maintenance expenditures running much above the average for similar property. Conversely, it may indicate below-average costs, suggesting the possibility of deferred maintenance and increased deterioration.

Physical deterioration is usually estimated as a percentage; a new asset has 0 percent deterioration, while an asset which is completely used up has 100 percent deterioration, with all other assets falling between those two extremes. Theoretically, physical deterioration can be measured objectively. A machine will produce x number of parts in its physical life. Assuming that adequate statistics were kept, the machine was never rebuilt or abused, and all assets of the same type are equivalent, then a simple ratio of the past production to the total production expected would be an objective measure of physical deterioration. Obviously machines are rebuilt, they are used and abused, and quality varies. Unless you are dealing with large assets, production statistics are usually not maintained for individual assets. Therefore measuring physical deterioration is subjective, so the appraiser must rely on how similar assets have performed in the past in order to make judgments of the physical condition of the subject.

Measuring Physical Deterioration. The best rule to follow when measuring physical deterioration is to rely on the facts and circumstances applicable to the subject, particularly the age and use of the asset. The machinery and equipment appraiser should strive to carefully segment the three elements of depreciation—physical, functional, and economic—and should deal with each element individually. When referring to physical deterioration, the appraiser should be certain that it is only the physical wear and tear that is being measured and not some overall depreciation representing physical deterioration plus some factor for functional loss or lack of utility. For example, a slide rule that is still in its original case, wrapped in plastic, and never used, has 0 percent physical deterioration. It is, however, 100 percent functionally and economically obsolete for the use for which is was designed and built.

The use of an asset is probably the best indicator of physical deterioration. Machines used 24 hours per day will physically deteriorate faster than the identical unit operated 8 hours per day. Machinery employed in dusty, dirty, abrasive, and/or corrosive atmospheres will deteriorate faster than the same asset in a clean environment. Machines that have just been rebuilt are in better physical condition than those which need to be rebuilt. Be aware that a fresh coat of paint does not change the physical condition of an asset. It may suggest a high level of maintenance, but a new paint job may also be used to cover defects. The appraiser must go beyond casual observation to investigate any extraordinary care and maintenance or abnormal wear and tear.

When an asset such as a machine tool is rebuilt, its physical deterioration is partially corrected because the rebuilt machine is not new and some physical deterioration (incurable) exists that cannot be corrected. This difference can be measured by comparing the difference between the selling price of a new unit with that of one rebuilt by the factory.

In some industries, certain maintenance practices are performed which will aid the appraiser in measuring physical deterioration. In the petrochemical industry, ultrasonic inspection of wall thicknesses for assets such as vessels, tanks, columns, and reactors is common practice. Tank bottoms and tops are replaced more frequently than tank walls. In the basic industries, refractories in melting furnaces are replaced regularly, while the shell suffers very little wear. These types of practices should be investigated and used in estimating physical deterioration.

In estimating physical deterioration, the appraiser should strive for consistency. While opinions will differ as to the condition of the same asset, a uniform judgment by an appraiser allows others reviewing the work to see the asset from the appraiser's viewpoint. Also, the appraiser should not allow personal feelings to influence judgment. Amount of depreciation cannot vary with the appraiser's prejudices.

Ultimately the measure of physical deterioration boils down to a comparison between the condition of a subject and that of an identical new asset. There are three methods of measuring physical deterioration: observation, formula/ratio, and direct dollar measurement. The remaining text discussing physical deterioration presents some ideas as to how to measure deterioration objectively. We hope the text will provide some insight and stimulate thought, but ultimately it will point out that isolating and measuring just physical deterioration is subjective and difficult at best. Nevertheless machinery and equipment appraisers must deal with the problem and should strive to base their opinion on the facts and circumstances.

Observation. This concept is nothing more than a comparison based on the experience of looking at similar assets as compared to a new asset. The procedure involves actually observing those elements of wear and tear that can be seen and converting the observations to a percentage. It also involves discussions with knowledgeable plant personnel to determine the condition of those things which are not

Fair Market Value Concepts

obvious, such as internal corrosion on tanks. Based on the facts, the appraiser must express an opinion as a percentage to deduct against the replacement cost.

Formula/Ratio. This concept involves analysis based on an asset's use or life, and since use is the best indicator of physical deterioration, it is discussed first. In its simplest form, physical condition can be estimated by the ratio

$$\frac{\text{Use}}{\text{Total use}}$$

Given some unit of measure, this ratio measures the amount of use of an asset at a point in time compared with the total use expected from that asset. For example, assume that the normal physical life of a machine is 100,000 hours. If a specific machine has logged 40,000 hours, it is logical to conclude that the physical deterioration is approximately 40 percent (40,000 ÷ 100,000 × 100 = 40 percent). In the case of a new asset, total use is synonymous with the expected use which in this example is 100,000 hours. Purists will argue, and rightfully so, that this is a linear or straight-line depreciation. In other words at 50,000 hours, the machine would be depreciated 50 percent, at 60,000 hours, 60 percent, etc. A valid question arises when using this method when the machine has 100,000 hours and is still operating; the simple ratio implies 100 percent depreciation. In this case, *total use* is defined as the sum of the usage to date (100,000 hours) and the remaining useful life, measured in hours. If the machine has 100,000 hours and is expected to last another 25,000 hours, the physical deterioration would then be 80 percent developed as follows:

$$\frac{100,000}{100,000 + 25,000} = \frac{100,000}{125,000} \times 100 = 80 \text{ percent}$$

Now let us consider the situation where the machine has 80,000 hours, has just been rebuilt, and has an expected remaining life of 50,000 hours. Physical deterioration would then be developed as follows:

$$\text{Percent} = \frac{80,000}{80,000 + 50,000} \times 100 = 61.5 \text{ percent}$$

Rounded 60 percent

When production statistics are not known, the ratio of an asset's age to its life can also be used as an objective measure of physical deterioration. For example, if an asset is 1 year old and we expect a 20-year physical life, the ratio of 1:20, or 5 percent, is an indication of the physical deterioration.

For larger, older, and more complex property, the concept is expanded to consider the ratio

$$\frac{\text{Effective age}}{\text{Total life}} = \frac{\text{Effective age}}{\text{Effective age} + \text{Remaining life}}$$

Example 3
Determination of Effective Age

The problem is to determine the effective age of an asset being appraised in 1981. You obtain the original costs and acquisition dates. Your investigation reveals that the property was purchased new in 1971 and expenditures were made in 1974 and 1976 for additions. A major overhaul was done in 1979, effectively replacing some of the original equipment.

The first step is to develop the proper basis for comparison, which in this case is trended original cost. This is determined by applying the appropriate cost index (in this case 10 percent/year) to the original cost for each year, which is shown as follows:

Purchase date	Original cost	×	Cost index	=	Trended original cost
1971	$20,000		2.60		$52,000
1974	2,000		1.95		3,900
1976	2,500		1.61		4,025
1979	17,400		1.21		21,175
	$42,000				$81,100

If we consider the *actual age* (or chronological age) as the number of years that have elapsed since an asset's inception, *effective age* reflects the amount of use the property has experienced during its life. It is the age indicated by the condition of the asset. To illustrate, consider the machine that has recently been rebuilt with many new parts. If its chronological age is 20 years, its effective age is something less than that because it is in better condition as a result of the overhaul.

The denominator in the preceding equation represents the total life expectancy of the asset. Total life may have three definitions depending on the age and use of an asset: (1) for a new asset it is the life in years that can be expected, (2) for a used asset it is the sum of the effective age plus the expected remaining life, and (3) for an asset that has been removed from service it is the actual number of years the asset was in service.

An advantage of using the age/life technique is that effective age can often be calculated using the client's fixed asset records. The effective age can be determined by weighting the investment in an asset or a group of assets. The weighting needs to be done on some equitable basis and must consider additions or deletions over the asset's life. The procedure can be used for a single asset (if the records are sufficiently detailed) or, more commonly, for groups of assets.

Fair Market Value Concepts

The original cost of $42,000 and trended original cost of $81,000 are somewhat misleading because they include some redundant investment resulting from the overhaul in 1979. In other words, these costs double up on those assets that were replaced as part of the 1979 overhaul. For example, if a pump is replaced in 1979, the cost is probably included twice; as part of the original investment in 1971 and again in 1979. To adjust, you must delete the redundant investment. For this example, we restate the 1979 overhaul in 1971 dollars by "back-trending" as follows:

$$\$17,400 \times \frac{1.21}{2.60} = \$8097$$

Rounded $8100

The original cost and trended original cost figures are then restated less the redundant investment and are summarized in the following table.

Purchase date	Original cost	=	Trended original cost
1971	$11,900		$30,825
1974	2,000		3,900
1976	2,500		4,025
1979	17,400		21,175
	$33,800		$59,925

The next step is to age the investment. This is done by multiplying the trended original cost by the age of the investment as follows:

Purchase date	Trended original cost	×	Age of investment, years	=	Weighted investment
1971	$30,825		10		308,250
1974	3,900		7		27,300
1976	4,025		5		20,125
1979	21,175		2		42,350
Total	$59,925				$398,025

The last step is to determine the composite age. This is done by dividing the weighted investment by the trended cost.

$$\frac{\$389,250}{\$59,925} = 6.64 \text{ years}$$

The result of 6.64 years is a reasonable estimate of the effective age of the asset we are trying to appraise.

The problem presented in Example 3 has been simplified to illustrate the techniques and concepts. We used cost information as the equitable basis for comparison. There are other bases which may be appropriate.

For example, when trying to estimate the age of a building, an appraiser may wish to develop the effective age on the basis of building area or, for equipment, may consider estimating the effective age on a capacity basis.

There are some shortcuts which are not as accurate as the method described in Example 3. One of the techniques is to use the trended original cost information to develop a composite cost index and interpolate using the cost index. If we did that in Example 3, the composite cost index derived by dividing the trended original cost by the original cost is 1.77. Interpolating that cost index indicates a weighted investment date of approximately 1975, or 6 years.

Some appraisers use this technique by weighting the original cost based on age (i.e., original cost × age in years). If the technique is used in the previous example, the effective age is 5.3 years.

The reasons for the differences with these shortcuts reflect the implied assumptions underlying the weighting schedule. The procedures outlined in Example 3 are believed to be the most accurate because they actually measure the age of the investment on an equitable basis. Establishing a composite cost index and interpolating is not as accurate because of the interpolation process as well as the fact that there may be short-term aberrations in the cost index. The third technique (original cost × age) is the least accurate because using age as a basis implies a proportional relationship, thereby giving equal weight to all costs. This problem can be significant in periods of high inflation.

The advantage of using the age/life technique is that we can reasonably estimate the effective age of an asset assuming that the client's records are available. The problem with using the age/life technique however is actually estimating the physical remaining life of that asset. If some physical problems are known that will limit the physical life, a reasonable estimate can be made. In reality, however, the appraiser is predicting the future, and therein lies the problem.

Example 4
Determination of Physical Condition

You are appraising a process furnace which operates continuously 24 hours a day, seven days a week, and you are trying to estimate its physical condition. You talked to the operating and maintenance personnel and determined that the furnace has operated normally since it was installed approximately 12 years ago. In your discussion, you learned that there had been some patching on the flues and duct work approximately five years ago, and some pumps, piping, and other external equipment were replaced approximately two years ago. You also learned that the refractory will need to be replaced in approximately five years.

Fair Market Value Concepts

For this example, assume that you calculated the effective age in a manner similar to that shown in Example 3 and you found it to be eight years.

The next step would be to estimate the composite physical remaining life for the entire furnace. In subsequent discussions you learn that once the refractory is replaced the furnace can readily operate for another 15 years, so the physical remaining life of the structural members is 20 years. Regarding the other equipment, it is in relatively good condition at this time but the operating and maintenance personnel do not expect it to last as long as the structural members. They expect to perform some maintenance on this other equipment to extend its life. The composite physical remaining life for the entire furnace can then be developed as shown in the following table.

	Percent of investment	×	Estimated physical remaining life	=	Weighted remaining life
Refractory	30		5		1.5
Structural members	50		20		10.0
Other equipment	20		15		3.0
	100				14.5

The composite physical deterioration is simply the ratio of the effective age divided by the total life, which computes to 35.5 percent as follows:

$$\text{Percent physical} = \frac{8}{8 + 14.5} = \frac{8}{22.5} = 35.5 \text{ percent}$$

Keep in mind that this is an estimate of the physical condition of the furnace as it exists. It does not consider the cost of the additional work that would be necessary to extend the life beyond that which is currently anticipated.

Although the age/life technique is discussed for estimating physical deterioration, it can be applied to other elements of depreciation when supported by the facts. For example, assume that you are appraising some pollution abatement facilities and a law has been passed requiring the owner to abandon the facilities within three years and replace them with different facilities. Furthermore, assume the existing assets are five years old and they have a physical remaining life of 15 years. In this case physical deterioration would be 25 percent (5 ÷ 20 × 100 = 25 percent). Since this asset has only a three-year remaining economic life because of the mandate, the total (overall) depreciation (physical, functional, and economic) is 62.5 percent developed as follows:

$$\text{Total depreciation, percent} = \frac{5}{5 + 3} = 62.5 \text{ percent}$$

An additional illustration to calculate overall depreciation appears in Example 9. Again, it is necessary to use this concept in light of the facts and circumstances which affect the asset you are appraising.

In the broadest sense we are considering straight-line depreciation when using this concept. The analysis establishes the position of the subject being investigated at an appropriate point in time in its life cycle. When adjustments are made for observed conditions relating to past and future use, and observed condition, this analysis becomes a valid tool in the appraisal process.

What is discussed here is a relatively simple way to look at age, life, and use of an asset. When data are available, there are more sophisticated ways of measuring depreciation. For additional information the machinery and equipment appraiser is directed to *Engineering Valuation and Depreciation,* written by Anson Marston, Robley Winfrey, and Jean C. Hempstead and published by the Iowa State University Press in 1982. The book discusses various methods of estimating service lives, including the Iowa-type survivor curves, and other methods of measuring depreciation.

The age/life technique is useful for newer assets or assets in midlife. When an asset requires a significant expenditure to solve a physical problem or a particular component has a short physical remaining life requiring replacement, it is suggested that the appraiser use this knowledge in estimating physical deterioration. The technique for doing so is described below.

Direct Dollar Measurement. This concept involves measurement of the dollar amount of physical deterioration. It is applicable when specific elements have deteriorated and can economically be cured, commonly referred to as *curable physical deterioration.* As applied to machinery and equipment, these elements include such things as the replacement of major components or rebuilding costs. Examples of this concept include a replacement of a motor, sandblasting and painting of a tank, or rebuilding of a machine tool.

When using this concept, the appraiser should strive to segregate the curable elements from the incurable. The primary difference between the two is that the curable physical deterioration can economically be cured while the incurable portion cannot. By segregating these elements we are analyzing the asset in two parts. The advantage is that the portion which is curable can usually be estimated directly in dollars providing yet another clue in estimating physical deterioration. Those elements which are incurable must be depreciated based on the observation or the formula/ratio techniques described earlier. The sum of the curable and the incurable represents the total physical deterioration for the existing asset.

Earlier in the chapter we pointed out that the appraiser should be careful not to double up when measuring the various elements of depreciation. For example, when estimating physical deterioration, be certain that only physical deterioration is being measured and not some combination of physical, functional, or economic depreciation. This be-

Fair Market Value Concepts 95

comes critical when the "cost to cure" is a significant portion of total depreciation. The dollars spent may relate to elements of depreciation other than physical deterioration.

Example 5

Appraisal of Tank Farm

You are in an oil refinery and have been asked to appraise the tank farm. You have determined that the storage requirements of a modern refinery would be the same as those of a tank farm. The only difference is that a modern refinery would utilize fewer but larger individual tanks to meet those storage requirements. The replacement cost was developed in total based on the configuration of the modern refinery's storage requirements. You have concluded that the replacement cost is approximately 10 percent less than the reproduction cost for the equivalent storage capacity. For this example, then, the replacement cost for the individual tanks is simply the reproduction cost less 10 percent.

During your discussion with the maintenance personnel, you were made aware of the fact that a crude oil tank has a small leak in its bottom because of corrosion from saltwater, which is naturally present in this type of crude oil. The tank has been pumped out and is being prepared for maintenance. Preliminary indications are that the entire bottom and the first course of the sidewalls will be either patched or replaced, depending on the wall thickness. The planned expenditure is approximately $350,000, which includes the costs associated with removing the tank from service, cleaning, preparing a safe work environment to make the repairs, removal, and replacement of the corroded areas.

You are using the cost approach and have determined that the reproduction cost for this tank is $2,000,000 and its replacement cost is $1,800,000. Your next step will be to determine the physical deterioration.

In this situation, we have a physical problem that is curable, i.e., replacement of the tank bottom. The total physical deterioration of the tank should be deducted from reproduction cost, since this is the specific asset which we are appraising. The result, expressed as a percentage of reproduction cost, should be applied to replacement cost. In this way, we are measuring the actual physical deterioration and deducting it proportionally from replacement cost.

The first step is to deduct the cost to cure:

Reproduction cost	$2,000,000
Less curable physical deterioration	−350,000
Equals reproduction cost of incurable	$1,650,000

The amount remaining is the cost of the portion of the tank subject to incurable deterioration. It may be estimated using the observation or formula/ratio techniques described in the text.

Assume that the tank is 10 years old and we estimate at least another 20 years of physical life (remember that these areas of the tank would not be affected by the saltwater).

The total incurable physical deterioration is 33.3 percent, or $550,000:

$$\frac{10}{10 + 20} = 33.3 \text{ percent}$$

Reproduction cost of incurable portion	$1,650,000
Incurable physical deterioration, percent	33.3
	$ 550,000

The sum of curable and incurable is $900,000, and when divided by the reproduction cost suggests a composite physical deterioration of 45 percent.

$$\text{Composite physical deterioration} = \frac{\text{Curable + incurable}}{\text{Cost reproduction new}} = \frac{350 + 550}{2000} = 45 \text{ percent}$$

The composite physical deterioration of 45 percent is then properly applied against the cost of replacement.

The three methods to estimate physical deterioration are all valid given sufficient information and an appropriate amount of time in which to analyze the data. In theory you should consider all three methods when estimating physical deterioration, but in practice this is not always practical. Let the facts and circumstances dictate the appropriate method or methods to be used in your appraisal assignment.

Sources of Information for Physical Deterioration. The primary sources for information in determining physical deterioration include capital expenditures, both historical and future, production records, and discussions with maintenance and engineering personnel. From the historical capital expenditures, an appraiser will get an indication of how much money has been capitalized for a particular asset. The appraiser may see that substantial money has been spent recently in a particular asset, suggesting that it is in reasonably good condition. Perhaps very little money has been spent, which may be normal for that asset or suggest that the asset may need major maintenance. When analyzing future capital expenditures (generally available in something like a five-year plan) an appraiser can readily see where major expenditures are anticipated. These expenditures may suggest that an asset is approaching the end of its useful physical life (as well as other functional or economic problems). A review of production records will provide an indication of how particular assets have been used. If production is down, it may be because of some physical problem. If the production level is up for an extended period, it may suggest an increase in physical deterioration that will shorten its life or require major maintenance.

There are no hard and fast rules regarding what should be ascertained when reviewing records. Records should provide some clues and ultimately lead to the facts on which to base an opinion. After the records have been reviewed, spend time discussing conditions with various maintenance and engineering personnel. It is difficult at best for

Fair Market Value Concepts

an appraiser or any outsider to ascertain the condition of a facility without talking to those individuals who must operate and maintain the assets on a daily basis. From these conversations the appraiser should gain some clues and ultimately additional facts upon which to base an opinion.

A word of caution regarding future capital expenditures: when reviewing things like five-year plans, it is quite common to see total replacements of assets. Often these replacements are made for reasons other than physical condition, and the appraiser should strive to ascertain the reasons why those particular assets are being replaced. Some of those future capital expenditures may be for repair and maintenance of the existing property, and some of it may be earmarked for expansion. If a complete replacement is planned in the future, it may relate to its physical condition, or it may need to be done to solve some functional or economic problem. If this is the case, there is additional depreciation affecting the asset being appraised.

Functional Obsolescence. The next step in developing the cost approach is to consider functional obsolescence.

Functional obsolescence is the loss in value within the property as a result of the development of new technology.

In the preceding chapter the author drew a distinction between functional obsolescence and technological obsolescence, *functional obsolescence* being a loss in value resulting from capability characteristics between a new machine and the appraised machine and *technological obsolescence* being a loss in value resulting from the difference between design and materials of construction used in present-day machines compared with those used in the machine being appraised. Notice that in this chapter we are defining functional obsolescence as being a loss in value as a result of the development of new technology. At first glance, an appraiser may be tempted to conclude that we have a conflict, but in reality the conflict is not with the subject matter but rather there is a legitimate difference of opinion as to how machinery and equipment appraisers apply the concepts to measure the functional/technological aspects affecting value. The basic point here is that it is important to measure the various factors which contribute to this element of depreciation. The names that are given the various elements—functional, technological, or something else—are secondary provided that the definitions specify what considerations are included in the terminology used.

For the remainder of this discussion we use the term *functional obsolescence* to mean a loss in value resulting from new technology. This includes such things as changes in design, materials, or process resulting in overcapacity, inadequacy, excess construction, lack of utility, or excess variable operating costs.

Functional Obsolescence from Excess Capital Costs. Functional obsolescence resulting from excess capital costs has already been discussed. It results from improvements and changes in layout, design, materials, product flow, construction methods, and equipment size and mix; in effect, improvements which made the new technology more desirable. The difference between reproduction and replacement cost represents the amount of excess capital cost.

Functional Obsolescence from Excess Operating Costs. The second aspect to be considered is operating obsolescence. Oftentimes, as a result of new technology, it is not only cheaper to build or buy an asset, but it is cheaper to operate. Calculating operating obsolescence involves a comparison of the operating characteristics of the subject property to its modern equivalent. The existing asset or facility and its higher operating costs are compared with the reduced costs being achieved in a modern replacement. The result of the study is the basis for estimating a penalty for continued use of the existing property. The study involves the following steps:

1. Analyze the operating statements and statistics of the existing property.
2. Determine similar costs for a comparable modern replacement facility.
3. Review the operating cost differentials.
4. Apply those operating cost differentials to projected annual capacities to arrive at the total annual excess operating costs which will be incurred as a result of continued operation.
5. Reduce the total annual excess operating cost to reflect the impact of income taxes on the resulting incremental income.
6. Estimate the remaining economic life during which time the excess costs will continue to exist.
7. Capitalize (convert to present worth) the annual excess operating costs at an appropriate rate of return over the remaining life.

Operating obsolescence can be defined as the present worth of the future excess operating costs. The subject property will continue to incur an additional expense over and above that of a modern facility. The penalty will continue until either the problem is corrected, the assets wear out, or, in extreme cases, the company goes out of business.

The concept of present worth is basically the inverse of compound interest, or, stated another way, it says that a dollar today is worth more than a dollar in the future. If you invest 91 cents today at 10 percent interest, it will grow to $1 in one year. Then, the present worth of $1 to be received one year in the future is 91 cents today at 10 percent interest. If a person would like to receive $1 per year over the next two years, how much should be invested today? At 10 percent and considering each year separately, the investment should be 91 cents today for the first year and approximately 82.5 cents today to receive $1 in two years. So the present worth of $1 to be received in each of the next two years is 91 and 82.5 cents or $1.735 at 10 percent. This is the process of capitalization in its simplest form. It is the conversion of expected future periodic payments to present worth.

When capitalizing these excess operating costs, estimate the remaining life of the facility and select the appropriate capitalization rate. Remaining life estimates have been discussed in Example 4. The elements included in the capitalization rate are

Prevailing risk-free rate

Amount of risk

Inflation

In addition, the capitalization rate can apply on a pretax or aftertax basis. Normally, operating obsolescence is calculated on an aftertax basis since the excess operating cost incurred by the subject would be taxable income in a modern plant. It is suggested that the selection of a capitalization rate be done with the assistance of an appraiser knowledgeable in financial valuation techniques.

Operating obsolescence is independent of any physical deterioration existing at the property. It is a form of obsolescence which occurs on-site and results from technological advances. Areas where operating obsolescence should be investigated are:

1. Operating labor
2. Maintenance labor and materials
3. Operating supplies and chemicals
4. Energy and utilities
5. Production yields

Functional obsolescence and, particularly, operating obsolescence can occur in the following situations:

1. Plants involved in the process industries
2. Plants involved in industries that either use assets or manufacture products with a high degree of technology
3. Older plants that have increased in size over time

4. Plants in which there are a number of identical units
5. Plants involved in industries which handle large volumes of material
6. Plants with areas of inactive machinery

For this discussion, we consider for analysis only those direct costs associated with operating obsolescence including labor and energy costs: those costs which can be directly attributable to a change of difference in technology. Other costs may be considered in the analysis. Those are depreciation differentials and the ultimate effect on taxes—basic capital budgeting criteria. The detailed discussion of these concepts is beyond the scope of this book and should make for interesting reading in advanced texts.

Up to this point in the discussion of operating obsolescence, we have been doing nothing more than comparing the operating efficiency of a particular asset or plant to its modern equivalent. In effect, this procedure is a relatively simple income approach (in this case, capitalizing a negative income). The issue of capitalizing income is discussed here because many appraisers assume that operating obsolescence doubles up the functional penalties; the accusation is that an appraiser has measured all the functional obsolescence by estimating replacement cost. This is absolutely not true. The point, especially when dealing with the issue of technology, is that in many cases it is not only cheaper to purchase a new asset (capital cost) but it is also cheaper and more efficient to operate (operating cost). In the examples which follow, we illustrate these points.

Example 6
Functional Obsolescence

In Example 1, we discussed how to determine the replacement cost for four boilers which were part of a petrochemical facility in the Midwest. Now we consider the operation of those boilers and look at the operating obsolescence associated with them. In your discussions, you learned that a total of 19 people are required to operate the existing boiler facilities.

Boiler number	Capacity, lb/hr	People
1	15,000	0
2	30,000	4*
3	30,000	7
4	60,000	8
	Total	19

*Average over a one-year period.

Fair Market Value Concepts

In addition, you learned that it requires only 10 persons to operate the boiler facilities in the modern refinery. (Recognize in this example, as well as in practice, that the modern equivalent is far more desirable because it is cheaper to build and, from a labor standpoint, much cheaper to operate.) The nine extra people represent additional production costs for the subject property compared with its modern counterpart, which places the subject at a disadvantage. This disadvantage—the annual excess operating costs—is the number of excess employees multiplied by the total labor rate. You spoke with the client's accountant and determined that the direct labor rate for each of these people as of the appraisal date is $25,000 per year, and the benefits paid by the company (insurance, FICA, and other benefits) amount to 30 percent of the base rate. Therefore, the total cost per person is $32,500 per year, which equals $292,500 per year for the nine people. This represents the annual excess operating cost associated with the existing facility.

The next step is to convert these excess operating costs to present worth. For this example, we assume:

1. A reasonable life expectancy of 10 years for the boilers
2. A 50 percent income tax rate
3. A 10 percent capitalization rate

Operating obsolescence is developed as follows:

Annual excess operating costs	$ 292,500
Less taxes at 50 percent	−146,250
Annual excess operating costs after tax	146,250
Present worth factor of 10 percent for 10 years	× 6.145
Operating obsolescence (labor)	$ 898,700
Rounded	$ 900,000

This is a relatively simplified example, but in practice it requires a substantial amount of experience and research to measure the operating cost differentials objectively. The point is that if an individual asset or an entire plant produces products at a greater cost, the assets are less desirable.

Example 7
Operating Obsolescence

In Example 2, a case was discussed where the replacement cost can be greater than the reproduction cost. The primary reason for this was that an advancement in technology has increased the capital cost but reduced the operating costs to produce each unit. In that example, the capacity is equal to 100 units per day, and the subject required 100,000 therms/year, whereas Model B required only 75,000, a difference of 25,000 therms. At 50 cents/therm, this indicates an annual operating cost savings of $12,500 per year.

Assume you have analyzed the physical condition of the asset and have concluded its value is 40 percent by estimating its effective age of six years and concluding that its normal physical life expectancy is approximately 15 years. Based on the information you have obtained, you have no reason to conclude that the normal life will be shorter or greater than 15 years, which implies a remaining physical life of nine years.

The next step is to determine the period of time over which you capitalize these cost savings: the economic remaining life of these assets. You discussed the situation with the appropriate operating personnel who agree with you that a nine-year physical remaining life is a reasonable estimate. The operating personnel agree that at the end of its physical life, major expenditures will be required not only to improve its physical condition but also to improve its operating performance.

The operating personnel agree that because of the advancement of technology their asset is less desirable from an operating standpoint. Nevertheless, they have no plans to replace this asset in the immediate future. Based on these facts then, you conclude that this operating penalty will continue for the remainder of this asset's physical life, nine years.

Assuming a 10 percent capitalization rate and a 50 percent tax rate, the development of operating obsolescence is as follows:

Annual excess operating costs	$12,500
Less taxes at 50 percent	−6,250
Annual excess operating costs after tax	6,250
Present worth factor of 10 percent for 9 years	5,759
Operating obsolescence: energy	$35,993
Rounded	$36,000

The conclusion of value by the cost approach is $60,000, which is developed as follows:

Cost of reproduction new	$130,000
Cost of replacement	160,000
Less physical deterioration: 40 percent	−64,000
Equals cost of replacement less physical deterioration	96,000
Less functional obsolescence from excess operating costs	−36,000
Equals cost of replacement less physical deterioration and functional obsolescence	60,000
Less economic obsolescence	−0
Equals fair market value-in-place	$ 60,000

There are two items that need to be discussed regarding the previous example. The capitalization process assumed that the annual excess op-

Fair Market Value Concepts **103**

erating cost of $12,500 per year is constant; i.e., production and the cost of energy will remain constant tending to understate the penalty. In theory, the capitalization rate should reflect the assumptions used in projecting the operating cost. If, in Example 7, the operating cost projection included increased production as well as inflationary increases in the cost of energy, this would be a riskier projection requiring a higher capitalization rate.

There can be a problem when analyzing operating costs for facilities operating at significantly reduced capacities. Those elements which contribute to operating obsolescence are known as *variable operating costs;* they vary with production. To produce a given product, x amount of energy, y amount of labor, and z amount of raw material are required. At capacity there is a linear (direct) relationship between the amount of input and the amount of output. Generally, the relationship holds true to about 75 to 85 percent of operating capability, but below those levels the relationship is not linear for all elements, which results in greater operating costs per unit of output. The discussion of this concept is beyond the scope of this book, but realize that distortions can and do occur.

Example 7 reflects the fact that additional capital costs are offset by reduced operating costs. There were sufficient data to draw an objective conclusion of value using the cost approach for that particular asset. In some cases there may be a dramatic cost increase (beyond inflation) without any corresponding offset for operating cost savings. This is especially true when appraising such things as pollution abatement facilities when there has been a change in regulations. For example, consider an appraisal of a dust collection system that was designed for particulate emission of 100 ppm (parts per million), and the facility operates within its designed capability. Assume that the government has passed a regulation that says all new dust collection systems can only emit 50 ppm, but this requirement effectively doubles the capital cost and in fact will require greater energy and labor to meet the requirement. Furthermore, assume that the existing dust collection systems, including the one you are appraising, will remain in operation under a grandfather clause in the regulation. The question here is: What is the proper level of current cost for the dust collection system you are appraising? The answer is that the proper level of current cost is the reproduction cost because the specifications for the modern facility are far greater than those for the facility which you are appraising. In fact, in this example, there is no modern equivalent because of the regulations. In addition, a prudent investor is not going to build the newer dust collection system without realizing some economic benefit (reduced costs and/or improved quality, for example) or without being mandated, which is the case in this instance. The situation in which there is a substantial increase in capital cost not offset by any economic benefit is known as *betterment*.

To summarize: we have taken a brief look at operating obsolescence. Most appraisers understand and agree on the concept, but they often disagree regarding the details. Areas of disagreement may include use of an after-tax or pretax penalty, method of determining the capitalization rate, and the inclusion or exclusion of depreciation differentials and the effect on taxes. Further discussion of these topics properly belongs in advanced texts.

Much has been written on the subject of capitalization rates. For a theoretical background, it is suggested the reader review finance textbooks. As applied to the appraisal process, two books are recommended: *Valuing Small Businesses and Professional Practices*, written by Shannon P. Pratt and published by Dow-Jones/Irwin, 1986; and *The Appraisal of Real Estate*, 8th edition, published by The American Institute of Real Estate Appraisers, 1983. In addition, there are numerous articles published on the subject in various periodicals such as *Valuation* (American Society of Appraisers).

Economic Obsolescence. The last step of the cost approach is to estimate economic obsolescence.

Economic obsolescence is the loss in value resulting from factors external to the property.

The causes of economic obsolescence may include reduced demand for the product; increased competition; changes in raw material supplies; increasing costs of raw materials, labor, or utilities without a corresponding price increase of the product; inflation; high interest rates; legislation; and environmental considerations.

Measuring the full effect of economic obsolescence is the primary weakness of the cost approach. This is because economic obsolescence is a function of outside influences which affect the entire business (all tangible and intangible assets) rather than individual assets or isolated groups of assets. Economic obsolescence is best measured through the use of the income approach, but its use in the cost approach is discussed in the following section.

Inutility. There is a good way to measure certain forms of depreciation using a derivative of the cost-to-capacity relationship discussed elsewhere in this text (see Appendix at the end of this chapter). Whenever the operating level of an asset or an entire plant is less than its rated capability, an inutility penalty exists. The penalty reduces the capital investment from rated capability to the actual operating levels to "balance" the plant. For example, you are appraising an asset that has a rated capacity of 1000 tons/day and is operating at only 600 tons/day. If you develop your replacement cost on the basis of 1000 tons/day and capitalize the operating cost penalties at 600 tons/day, an obvious imbalance exists. That imbalance is the additional unproductive capacity reflected in the capital cost estimate but not reflected in the operating obsolescence. This unproductive capacity should be reflected in your depreciation estimate.

The inutility penalty is calculated on a percentage basis by comparing the actual operating level to the rated capability through the use of the following formula:

$$\text{Inutility, percent} = \left[1 - \left(\frac{\text{Capacity B}}{\text{Capacity A}}\right)^n\right] \times 100$$

where capacity A = rated capacity
capacity B = actual production
n = scale factor

This relationship is based on a cost estimating technique whereby the cost of facilities of different capacities varies exponentially rather than linearly because of the economies of scale. In other words, as capacity increases, cost also increases but at a different rate. This same logic is used to develop the inutility penalty. Scale factors vary depending on the type of equipment and labor/material ratios. The factors range from 0.4 to slightly greater than 1.0.

It is important to note that use of this penalty applies both to functional and economic penalties. Again, its primary purpose is to balance the plant from both a capital and operating cost basis. Continuing with the example mentioned earlier, let's say you capitalize the operating obsolescence on the basis of 600 tons per day. The actual inutility is the 400 tons per day, which is not operating. Under the principle of substitution, a prudent investor would not purchase this unproductive capacity without being able to realize some benefit. If the plant is not operating at capacity for economic reasons, the inutility then becomes an economic penalty. If there is an imbalance in the productive capacity (bottlenecks), then the inutility penalty is probably functional obsolescence. Finally, although it is not common, it is possible that if a plant is not operating at capacity because of some physical reasons, then the inutility may be the result of physical deterioration. Once again, determine the facts and circumstances and apply them as appropriate. Now we consider some examples.

Example 8
Economic Obsolescence

You are appraising a production line capable of 1000 units per day. It is approximately three years old and is in excellent condition. Technically it represents the state of the art. In your discussions with the client, you have learned that there has been a dramatic increase in foreign competition for this product. As a result, the client is operating the facility at only 750 units per day.

For this example, let us assume that the replacement cost is $1,000,000 for 1000 units per day capacity and that the physical deterioration is approximately 15 percent. Your charge is then to determine the additional depreciation as part of the cost approach.

The reduced operating level is an element of economic obsolescence since it is caused by factors external to the property. In this case, an economic inutility penalty is appropriate and is developed as follows:

$$\text{Inutility, percent} = \left[1 - \left(\frac{750}{1000}\right)^{0.7}\right] \times 100$$

$$(1 - 0.818) \times 100$$
$$= 0.182 \times 100$$
$$= 18.2 \text{ percent}$$

The fair market value then for this asset developed by the cost approach is summarized as follows:

Cost of replacement	1,000,000
Less physical deterioration (at 15 percent)	−150,000
Equals cost of replacement less physical deterioration	850,000
Less functional obsolescence from excess operating costs	−0
Equals cost of replacement less physical deterioration and functional obsolescence	850,000
Less economic obsolescence (at 18.2 percent)	−154,700
Equals fair market value in place	$ 695,300
Rounded to	$ 700,000

In the preceding example there are a number of things worth discussing. First, notice that the inutility penalty is not linear; a 25 percent decrease in operating capacity results in an 18.2 percent inutility. Second, the penalty was taken after physical deterioration and functional obsolescence even though there was none; economic obsolescence is independent of physical deterioration and functional obsolescence. Third, a valid question can be raised regarding the proper capacity to use for developing value by the cost approach. If the economic condi-

Fair Market Value Concepts   **107**

tions are such that the long-term production will be 750 units per day, it may be valid to calculate replacement cost on that basis. Again, determine the facts and circumstances and apply them accordingly.

Using an inutility penalty is a way of measuring one aspect of economic obsolescence within the cost approach. In practice, when dealing with relatively new assets that are not operating at their capacity because of economic reasons, additional economic obsolescence is probably present. To measure this requires a detailed analysis of the business and a subsequent allocation of any economic penalties to the individual assets or groups of assets.

Overall Depreciation

Earlier in the text the age/life technique was described to estimate physical deterioration. We also mentioned that it could be used if certain facts and circumstances existed to measure an overall (or economic) depreciation. This use is now illustrated.

Example 9
Age/Life Technique

Assume we are appraising the same machine as discussed in Example 7. The conclusion of value was $60,000, developed by determining the replacement cost of the modern equivalent, estimating physical deterioration, and capitalizing the excess operating costs to arrive at operating obsolescence.

There is enough information in that example to use the age/life technique to calculate a composite overall depreciation. The effective age was 6 years, and the remaining physical life was 9. You concluded that the asset would continue to operate for the remainder of its physical life; hence the total economic life is 15 years (6 + 9 = 15). The overall depreciation then is the ratio of 6 to 15, or 40 percent.

Notice that the physical deterioration percentage in Example 7 is equal to the overall depreciation percentage in this example, 40 percent. The reason that the numbers are the same is because we have concluded that the physical remaining life is equal to the economic remaining life. The key difference between the two examples, though, is the application of that percentage to the appropriate cost level.

In Example 7, and in keeping with the principle of substitution, we were comparing the asset we were appraising to the modern equivalent. All depreciation was deducted from replacement cost, the upper limit of value. In this example, using the relationship between age and life, we are measuring depreciation in terms of service life or utility for the specific asset being appraised (we are not comparing this asset to its modern equivalent). Accordingly, the overall depreciation of 40 percent derived in this example is applicable to the cost of reproduction.

The value for the particular asset is $78,000 developed as follows:

Cost of reproduction	$130,000
Less overall depreciation, 40 percent	−52,000
Fair market value-in-place	$ 78,000

Two different methods of using the cost approach have now been illustrated to appraise a single asset. The value conclusions differ by $18,000. Both techniques are reasonable given the assumptions made in each case. The final conclusion then would be based upon that approach which is most supportable.

It is our opinion that the method presented in Example 7 is the proper method to measure the fair market value-in-place because it more reasonably reflects how the asset is being used. Use of the age/life technique reflects only that the asset will be used but cannot measure the effectiveness of how it is being used.

One final comment regarding the use of the age/life technique. Once again, what is presented here is a simplified version of the concepts behind the development of the Iowa curves. In effect, the age/life method of depreciation measures the amount of service life for utility that has been exhausted and by deducting that from the total anticipated service life results in the service life remaining for that asset. The underlying assumption is that each year's service life is worth a proportional amount; the first year's service is worth the same as the last year's service. This is simply not true if we consider the fact that a machine in the last year of its life has suffered from physical deterioration and functional and economic obsolescence; its ability to contribute in its last year is certainly something less than its ability when it was new. *Engineering Valuation and Depreciation* by A. Marston, R. Winfrey, and J. Hempstead (Iowa State University Press, 1968) presents some depreciation methods based on interest theories (present worth) to overcome this deficiency in the age/life technique. This concept suggests that this year's service provided by an asset is worth more than the same service provided in future years. Use of these methods provides a more accurate and more realistic picture of the overall depreciation (based on age/life), but they are relatively complex to apply and assume that the appraiser has adequate data with which to work. For more information refer to *Engineering Valuation and Depreciation*.

Sequence of Depreciation

We have discussed the use of the cost approach using a very definite order to estimate depreciation: current cost less physical deterioration, functional obsolescence, and economic obsolescence. This is the generally accepted sequence of deducting these various elements of depreciation from current cost. The logic of the sequence is derived from a normal life cycle of an asset. When an asset is new, its value is equal to the price at which the item actually sells. There is a willing buyer and a willing seller, and the implied assumption here is that there is an economic justification (a business need of some kind) for the purchase of this asset. Once the asset leaves the manufacturer, it begins to depreciate. Usually the first element of depreciation that occurs is physical deterioration, since the asset will probably be placed in service and used for the purpose for which it was purchased. As the asset continues in operation, two elements of deterioration come into play—curable and incurable. The curable is in the form of routine maintenance, and the incurable begins in the form of such things as metal fatigue. Physical deterioration is the only element of depreciation which occurs until something happens in the marketplace or environment to trigger functional or economic obsolescence. Usually a manufacturer will improve a product gradually over time, so when a manufacturer announces a "new and improved" version of a machine, obsolescence is usually introduced into the existing assets. Normally the new version is a result of some technological improvements, suggesting some functional obsolescence. When there is a dramatic change in technology, functional obsolescence can be significant. At this point the asset is operating, it is physically deteriorating, and now it exhibits some functional obsolescence. As time goes on, external factors such as reduced profitability in an industry, increased competition, foreign imports, a shift in market demand or government regulations lead to economic obsolescence. This then is usually the last element of depreciation to affect an asset.

This is the normal sequence of depreciation when using the cost approach. The sequence may vary when the facts dictate. The point, though, is that when using the cost approach an appraiser should strive to segregate the various elements of depreciation and make sure not to double up on depreciation or leave something out.

Summary: Cost Approach

We have used the cost approach to develop fair market value-in-place. The question is, "Is this fair market value?" Some machinery and equipment appraisers will look at the detail on the cost approach and say that they have all the elements of depreciation and conclude, "Yes, it is fair market value." The key to answering the question as to whether fair market value-in-place is actually equal to fair market value depends on whether we have the full amount of economic obsolescence, which, we will see, is the biggest weakness of the cost approach.

The most effective way to measure the full effect of economic obsolescence is through an analysis of the earning capability of the entire business in which the assets are employed. This earnings analysis is beyond the operating cost level. It considers an analysis of all revenue and costs associated with the business. The elements of depreciation and physical, functional, and economic obsolescence discussed so far are the result of analyses of assets. Another analysis is required to see if the business can produce earnings to generate a satisfactory return on all its assets, tangible and intangible. The results of that analysis will determine if, in fact, there is additional economic obsolescence attributable to the machinery and equipment.

So, under the premise of continued use, the answer to the question is:

- Yes, if it is assumed that there are adequate earnings to support the level of value the appraiser has concluded.
- Yes, if in fact the appraiser has analyzed the earnings capability of the assets and concluded that the earnings level is sufficient to support that level of value.
- No, if the appraiser has done an earnings analysis and concluded that the level of value cannot be supported by the earnings capability, in which case additional economic obsolescence is present.

Market Approach

The logic behind the market approach under the premise of continued use is that a prudent investor can go to the marketplace and purchase an existing operating facility or purchase individual pieces of equipment in the used market to assemble an operating "package." The market approach is used in the used market to establish value by analysis of recent sales or asking prices of comparable property. For machinery and equipment, the used market is an established means of buying and selling equipment. That market consists of used machinery dealers, auctions, and public and private sales, and as such the used market is a good source for determining the value of equipment.

The market approach, also known as the sales comparison approach, as applied to machinery and equipment is defined as follows:

> The *market approach* is that approach to value where recent sales and offering prices of similar property are analyzed to arrive at an indication of the most probable selling price of the property being appraised.

The basic procedure is to gather data, determine the appropriate units of comparison, and apply the results to the subject.

In using the market approach, the machinery and equipment appraiser has a couple of alternatives available, although one is more practical than the other. The first alternative is to utilize the market approach by assembling an operating package: purchase individual assets in the used market; dismantle and move; and add tax, freight, rebuilding, installation, and connections costs. The second method is to compare the subject property on a total plant basis to sales of similar type properties which have recently changed hands. In other words, it is possible to appraise an operating plant by comparing it to sales of other operating facilities. Unfortunately, this is impractical.

It is rare when only machinery and equipment are sold for continued use. Usually any total plant sales involve not only machinery and equipment but real estate and intangibles. In addition, a major factor affecting comparability is the product being made and how much income can be generated from the sale of that product, which will affect fair market value for continued use. Finally, when entire plants are sold in liquidation it is difficult, if not impossible, to compare the sale of a "dead plant" to the sale of a facility that continues to operate. For example, at the time this book was first written, the United States Steel Industry was experiencing one of the most difficult economic times in its history. A number of obsolete steel mills were being shut down and offered for sale in the marketplace, and some were sold. It is incorrect to compare the sale of idle facilities to similar facilities that are operating without a thorough analysis of the industry and an analysis of the earnings potential of the subject property. Because of these factors, the use of the market approach as applied to total properties will not be discussed. Rather, this discussion concentrates on the market approach as it applies to individual items of machinery and equipment.

Elements of Comparability

Ideally, when appraising machinery and equipment using the market approach, the appraiser should strive to base conclusions on sales of identical assets which have been exchanged in the marketplace. Unfortunately, it is rare when we have multiple sales of identical units. In practice, the market investigation will probably reveal sales of similar assets, and it is the analysis of similarity upon which the appraiser should base an opinion of value. The following text will identify and briefly discuss some of those items of comparability which pertain to machinery and equipment.

Age of unit. The appraiser should strive to compare the subject to sales of assets of the same vintage.

Condition. This is a difficult area of comparison because while the condition of your subject is known, it is difficult to know the condition of the comparable. Differences in condition may justify variances in selling prices of similar assets. An investigation of condition of the comparables should be done, if possible.

Features (accessories). The appraiser should strive to compare the same features and accessories.

Location. The geographical location where the sale took place will definitely affect the selling price. In addition, the physical location of an asset within a plant will also have bearing on the sales price. Two identical package boilers—one on the main floor and one on the third floor—will have different selling prices because of the difference in dismantling and moving costs.

Manufacturer. The appraiser should strive to compare sales of similar assets made by the same manufacturer.

Market conditions. This is probably the most important factor in appraising machinery and equipment and one of the most difficult to measure. The question is, How will the market affect the value of my subject? In most cases, a declining or depressed market may be a basis for discounting, while an increasing or inclining market may be a basis for enhanced values. Supply and demand should be considered. Is it a buyer's or seller's market?

Motivation. This is a key item of comparison, especially for larger units. The questions here are "What is the motivation of the buyer?" How will it affect the value of the subject? For example, in most cases there will be a difference in the sales price if an asset is purchased by a dealer for speculation (a liquidation premise) or purchased by an end user (a continued use premise).

Price. In many cases, especially on larger properties, the transaction price should be investigated and expressed on a cash basis. This is particularly true if any favorable financing is involved.

Quality. The quality of your comparables should be equivalent to that of your subject. A difference in quality may be a basis for adjustment.

Quantity. Unit prices will vary considerably depending on the quantity. It is improper to compare the unit price derived from the sale of 100 identical automobiles to appraise just one. Quantity is also related to market conditions; a buyer's market suggests that a larger quantity is available, while a seller's market suggests a limited quantity.

Size/type. Ideally, the same size and type of equipment should be used for the comparison.

Time of sale. The appraiser should strive to compare sales which have occurred within a reasonable amount of time of the appraisal date.

Type of sale. The type of sale and the terms of the sale generally indicate different price levels. The same asset that is purchased by a machinery dealer at an auction (a liquidation premise) will probably have a higher price when it is sold by the dealer to an end user (a continued-use premise).

These elements of comparison have been presented in alphabetical order. The importance of each depends on the data available and the appraisal assignment.

Market Approach Techniques

The following list outlines three techniques for establishing value of individual items of machinery and equipment using the market approach.

Direct match. This technique establishes value on the basis of a direct match to an identical asset. Perhaps the best example is estimating the value (either retail or wholesale) of automobiles using a bluebook. If the manufacturer, the model number, the age, and the accessories are known, it is simple to find the value of the subject in the various automobile bluebooks. Adjustments are limited to mileage and, more importantly, condition. In this case the appraiser is directly comparing the subject to the compilation of sales of other identical autos. The direct match concept is certainly nothing new, but it is identified as a separate technique because it provides what is probably the most accurate indication of value using the market approach. If you do not have a direct match, your conclusions become more subjective.

Comparable match. This concept involves the development of value based on analysis of similar (but not identical) properties using some measure of utility (size, capacity, etc.) as the basis of comparison. For example, when appraising an engine lathe manufactured by Company A, the appraiser has comparables of other similar engine lathes of the same size manufactured by Companies B and C. Obviously, compared to a direct match, this technique becomes more subjective, which requires additional analysis of those elements of comparison discussed in the preceding section.

Percent of cost. This technique is nothing more than establishing the ratio of the sales price to the current cost new of an asset at the time of sale. With sufficient data, similar property can be analyzed statistically and relationships developed between age, selling price, and cost. For example, an appraiser is investigating a 16-by-108-in. engine lathe manufactured by Company A. In the market investigation, a direct match is not found, but many similar lathes manufactured by different companies are found, including Company A. The sizes are either much smaller or much larger than the subject. If the analysis suggests that the sales of engine lathes with an age similar to the subject are in the range of 40 to 50 percent of current cost, it is logical then to conclude value by determining the current cost of the subject and applying the appropriate percentage.

An important point is that regardless of the technique used, the more information an appraiser has regarding sales of similar assets, the more supportable the conclusions.

Determining Value for Continued Use

To this point, the value of individual assets has been discussed without regard for how those assets are being used. To use the market approach to appraise machinery and equipment under the premise of continued use, those elements which add value-in-use must be identified and included in the appraisal. In effect, the appraiser converts the market price of the base unit to fair market value-in place. For machinery and equipment, these elements include such things as freight, installation, connections, foundations, rebuilding costs, and any indirect costs such as engineering or design fees required to place an asset in service (the same costs considered in the cost approach).

To determine value for continued use using the market approach, the appraiser in effect assembles an operating package, purchases individual assets in the used market, dismantles and moves them if necessary, brings freight to the site, incurs rebuilding costs if necessary, and installs the unit. Given a value of an individual asset by the market approach, the next step is to convert the value of the uninstalled unit to fair market value-in-place.

Consider the situation where a used machine is purchased from a machinery and equipment dealer. Assuming a cash transaction, the buyer will pay freight from the dealer to the loading dock, and the machine

Fair Market Value Concepts   **115**

comes with a 30-day warranty on parts and labor. The additional expenditures required by the buyer to make this an operable unit contributing to the operation are freight; installation, including unloading, moving, and setting in place; foundations and millwright work if applicable; connections, including piping, wiring, and instrumentation; and start-up and testing. At this point, a used machine with a 30-day warranty is installed and operating and represents the market price for the machine and current cost to make that machine operable.

There is a major difference of opinion regarding the depreciation of those costs incurred to make that used machine operable. The philosophies regarding the depreciation of those additional expenditures are:

1. All additional expenditures are depreciated using the same techniques as described in the cost approach.
2. None of the expenditures are depreciated.

Both philosophies have merit depending on the application. In Philosophy 1, all additional expenditures are depreciated, and it is applicable in most appraisal applications given certain assumptions. For example, when appraising an asset that was purchased new and installed 10 years ago, it would be logical to estimate the value of the base unit from comparables which have been exchanged in the marketplace. Because the subject is operating and has been operating for some time, obviously the additional expenditures (freight, installation, and connections, etc.) have depreciated over time and the appraiser is obligated to recognize that loss in value. The fact is that the asset including its connections and installation is not new.

Philosophy 2 suggests that none of the additional expenditures are depreciated but with certain assumptions this idea has merit. For example, assume a person has been asked to appraise some operating machinery and equipment for insurance purposes, assuming that there is a total loss, and the appraiser is to go to the used equipment market to replace the assets and reinstall them at the site. In this situation the *value for continued use* would be the sum of the market prices for the individual assets plus the undepreciated cost to place those assets in service. The result reflects used machines that have been newly installed.

To summarize, keep in mind that we are appraising operating assets by comparing them to transactions in the marketplace. We are not appraising the comparables but rather using them to provide an indication of value of the asset being appraised. We are then using the cost approach to measure the value of those items which convert the base unit from a market price to fair market value in place.

Consider a second situation in which a person purchases an asset in liquidation and intends to use that asset in an operation. Assume that the person paid cash for this asset and purchased it on an "as-is, where-is" basis. In this case the buyer is responsible for the dismantling and removal of this asset as well as any maintenance or rebuilding expenses. These costs incurred to the buyer would normally be "buried" in a dealer selling price, if the asset were purchased from a dealer. In theory the total consideration (purchase price plus dismantling, removal, rebuilding, and/or maintenance expenses) for a dealer should be the same as that for an end user with the exception of the additional costs for the dealer's overheads and profits.

Example 10
Valuation Using Market Approach

You have been asked to estimate the fair market value in place for a milling machine using the market approach. The machine is currently being used for custom work and is used frequently. It is eight years old and maintained on a preventive basis. Your observation is confirmed by discussions with plant personnel that the machine is in good condition.

Since you attend auctions regularly, you know that this is a very popular machine. Your data suggest a range in selling prices from $1000 to $7500, depending on age and condition. Recent sales suggests the subject would bring between $5500 and $6500. You talk to a used machinery dealer and describe the subject. The dealer says he or she would ask $6500 for this asset and probably sell it at $6000.

Based on your knowledge of the market and the confirming discussion with the dealer, it is logical to conclude that $6000 would represent the value of the base unit. You must now add those items which convert this base unit to fair market value-in-place.

Your knowledge of this asset indicates that this is a relatively simple item to install and connect. Because this is a common item, you assume that the asset would be purchased locally and estimate the freight cost to be approximately $200. You estimate that two millwrights could unload and set the machine in place in approximately two hours for a total cost of $125. Similarly you estimate the electrical installation including controls at $300. The total cost (new) is $625.

Since the asset was installed when new, all these in-use elements are eight years old and accordingly should be depreciated. They are not new. Based on an age of eight years and an expected life of 20, you estimate depreciation to be 40 percent ($8 \div 20 \times 100 = 40$ percent) and when applied to the $625 results in the value of the connections equaling $375. The sum of the base unit plus the value of the connections then is the fair market value-in-place of the subject which, in this case, is $6000 plus $375, or $6375.

The example has been simplified to illustrate the concept behind the market approach. The exact numbers used are, of course, fictitious, but a range of selling prices from $1000 to $7500 is realistic. To conclude, a specific number from that broad a range requires a strong knowledge of the marketplace and supporting data; the better the data, the better the answer. Once again the conclusion should be based on the facts and circumstances unique to the situation.

Example 11
Value-in-Use versus Value-in-Exchange

You have been asked to estimate the value-in-use for a coal preparation plant. The subject is 10 years old and rated at 700 tons/day. Maintenance is performed on a preventative basis, and the plant is in reasonably good condition. Your investigation has revealed that there has been a recent upturn in the coal mining industry, and because of this, there seems to be an increased demand for coal preparation plants. You have already concluded value by the cost approach, and during your investigation the client made you aware of two sales of coal preparation plants within the last year. You discuss the situation with your client, who authorizes additional time for you to investigate these sales and possibly conclude value by the market approach.

The first sale involved an exchange of ownership of an operating company. The assets of the company included the coal preparation plant, many trucks, other above-ground mining equipment, and two long-term contracts negotiated prior to the sale of the company. The company's primary business is to process coal for two nearby strip mines.

In your discussions you have heard that the purchase price was approximately $10,000,000. You call the new owner who says that the published purchased price is "a little high" but does not share any detailed information with you. The new owner does say, however, that he "wanted the contracts." You conclude that this is not a valid sale to compare to your subject since this was the sale of an operating company and included other assets, and in addition you had no facts for comparison.

Sale 2 involved the sale of a coal preparation plant rated at 1000 tons/day, which has been idle for the last two years. The former owner declared bankruptcy in the midst of the recession. The bank retained ownership and contracted with a local equipment broker to maintain the facility and subsequently sell it in operating condition. It was recently purchased by a major coal company who moved it to operate at a nearby site.

You talked to the dealer who sold the plant. The dealer confirmed that the bank held the asset, hoping that the market would rebound, and in fact, it did. He said there was a substantial amount of interest from many companies, but the closest buyer bought the plant.

The plant was eight years old and required minimal capital expenditures since the broker operated the plant periodically to ensure that everything worked. You contact the buyer who says that they paid $3.5 million for the plant and spent an additional $1.5 million to move it and install it at the site. The buyer is very satisfied and he feels he got a good deal. You pursue the "good deal" comment further, and find that the reason they purchased this plant was because a new one of the same capacity is approximately $12 million installed. The buying company was going to operate this plant at a mine with a limited life estimated to be five years, and the purchase of a new plant for that application could not be justified. When this used plant became available, the buyer grabbed it.

Since you have only one sale of a coal preparation plant, you must decide whether it is applicable. Because there was a substantial interest in this plant (and for illustration purposes), we assume that this is a valid sale. In this situation the percent-of-cost technique would be appropriate. The ratio of the $5 million purchase price plus installation cost to the current cost of $12 million is approximately 40 percent (5 ÷ 12 × 100 = 41.7 percent, rounded to 40 percent). The fair market value-in-place of the comparable is 40 percent of current cost.

The subject is less desirable because it is a smaller size, it is approximately two years older, it is not in as good condition as the plant which recently sold, and the installation is not new. Obviously, because of these reasons, something less than 40 percent of cost would be appropriate for the subject. For illustration purposes we assume that the conclusion is 30 percent of cost for the subject. If the cost of the subject is $10 million (remember it is smaller) then the fair market value-in-place is approximately $3 million.

Once the conclusion of value is reached by the market approach, it should be compared to the results indicated by the cost approach. If the results of the cost approach are substantially higher, the market approach may be used as a basis to derive additional depreciation (probably economic obsolescence). If the results are reasonably close, your value conclusions are probably reasonable. If the results of the cost approach are significantly lower, the value derived by the market approach may be a more reasonable answer reflecting additional desirability in the marketplace. Once again, the facts and circumstances should influence the answer.

Invariably, the question will be raised as to whether one sale makes a market. Usually the answer is no, but there are a couple of unique circumstances that may exist to cause you to analyze a single sale and use it in the appraisal process. One circumstance was discussed in Example 11 where there were only one or two recent sales. As it turned out, only one was comparable, and its use in the appraisal process has been discussed.

Another circumstance exists where there is no market; little or no property is exchanging hands. This is common in industries that suffer severe economic problems such as the steel industry in the United States mentioned earlier. There are many steel-making facilities that have been shut down and offered for sale in the market, but there are very few, if any, buyers who purchase the assets for continued operation. The point is that market inactivity, such as in the steel industry, suggests a significant amount of economic obsolescence which is applicable to those facilities that remain and continue in operation.

Notice also in Example 11 that the market data are based entirely on hearsay; no documentation exists verifying the information portrayed. An appraiser should attempt to verify market data as well as possible.

Fair Market Value Concepts 119

Summary: Market Approach

Finally, the question is raised as to whether the values concluded by the market approach are really fair market value. The selling prices of individual assets reflect the supply and demand and the desires of buyers and sellers in the marketplace. For individual assets, these selling prices reflect all elements of depreciation for that asset. But again, as with the cost approach, we need to look at how those assets are being used in the business in which they are employed. There are three answers which are the same as those identified for the cost approach. Those are:

- Yes, if adequate earnings are assumed to support the level of value concluded.
- Yes, if there are adequate earnings to support the level of value concluded.
- No, if there are insufficient earnings to support the level of value concluded, which means that additional economic obsolescence is present.

The Income Approach

The use of the income approach is not discussed in depth because it is beyond the scope of this book. However, some comments need to be made because, as we have mentioned previously, the cost and market approaches cannot measure the full effect of economic obsolescence. The primary purpose of this discussion is to illustrate to the machinery and equipment appraiser how we might derive this additional economic obsolescence if it is applicable when appraising machinery and equipment as part of an operating business under the premise of continued use. This section is not meant to imply that all machinery and equipment appraisers need to do financial analysis when performing their work. Rather, it is to provide additional insight into the appraisal process. The issue of valuing a business using an income approach and its relationship to machinery and equipment will be the subject of future publications.

In its simplest form, the income approach is the present worth of the future benefits (income) of ownership. It is not usually applied to individual items of equipment since it is difficult, if not impossible, to identify individual income streams. However, if you assemble a group of individual machines to produce a product, in aggregate, they generate income for the business. So by using an income approach, we can value the aggregation of assets which generate this income. This collection of assets is commonly known as the business enterprise and consists of all assets of the business—working capital and fixed and intangible assets.

The business enterprise is valued on the basis of its future income potential. Ultimately, buyers make decisions as to how much they can afford to pay for a business based on its future income potential and the related risk. The value of a business enterprise, then, is the present worth of the future income.

There are a variety of techniques used to value a business: discounted future earnings, direct capitalization, market multiple approach, and discounted cash flow, to name a few. All these approaches involve estimating some level of future income and converting that income to its present worth. Ultimately it is the income that establishes the value of the business which in turn determines the value of the individual assets being appraised under the premise of continued use. Through a series of calculations, the *amount supportable* by the business (derived from income) is determined for all the assets of the business including machinery and equipment. This amount supportable is compared to the fair market value-in-place to ascertain if any additional economic obsolescence is present.

If the business can support an amount equal to or greater than fair market value-in-place, the fair market value of the machinery and equipment is equal to fair market value-in-place and no additional economic obsolescence exists. If the business supports an amount less than fair market value-in-place, an additional amount of economic obsolescence is present. Then, fair market value-in-place must be reduced to the level supported by the business, which is then fair market value. In certain industries, especially those subject to severe economic problems, it is possible that the business supports a level of value equal to or less than liquidation value of the assets. When this situation occurs, fair market value is equal to liquidation value, and it may be prudent to sell the assets in liquidation rather than continue operating as an ongoing business.

We now consider a relatively simple example. An appraiser is asked to determine the fair market value of some machinery and equipment and is working with a business appraiser who asks the first appraiser to determine the fair market value-in-place and the liquidation value of the assets. They are $1 million and $2 million, respectively. The business itself is only worth $1 million on a going concern basis, and after some analysis the business appraiser concludes that the earnings of the business will support only $400,000 of value for the machinery and equipment. (The remainder is for real estate, working capital, and other assets.) This indicates that there is some additional economic obsolescence attributable to the machinery and equipment, which has not been measured through the traditional approaches. In this example, the fair market value-in-place must be reduced to reflect the additional penalty (additional economic obsolescence), so the fair market value of the machinery and equipment is $400,000.

Now, consider two other situations. One is if the income supports $1,200,000 for the machinery and equipment which has a fair market value-in-place of $1 million. This situation indicates to a business appraiser that intangible assets exist. If, on the other hand, the economic prospects of this business are so poor that the income can support only $100,000 for the machinery and equipment, then the fair market value equals liquidation value, or $200,000, because the most profitable use would be to sell the assets in liquidation rather than to continue in operation as a business.

The reasons for these differences boil down to the fact that we are appraising these assets as part of the business. We have valued the business based on its income potential reflecting the economic use of those assets. If the assets of the business are being used to their full potential (i.e., they are profitable) the assets tend to have a higher value. Conversely, if the assets are not being used to their fullest potential, they are less desirable. This fluctuation reflects those elements which cannot be fully measured using the traditional cost and market approaches which include such things as interest rates, inflation, competition, demand, market shifts, increased costs of operations, and reduced profitability. To properly determine fair market value of assets, these things must be considered and the best way is to consider them through the valuation of a business using an income approach.

The Need to Qualify Appraisals

To this point we have discussed the relationship between the value of assets and the business enterprise. We have stressed the importance of the income producing potential of the assets being appraised, and we have shown that the income (or lack thereof) affects the value of the fixed assets.

In the interest of maintaining standards within the appraisal profession, it is recommended that appraisals of machinery and equipment be qualified to address the issue of *economic supportability*. When appraising machinery and equipment an appraiser provides a better service to a client if the appraiser tells the client how the appraiser handled the issue of economics. To do this the appraiser needs to qualify the report.

If, for example, an appraiser has determined the fair market value-in-place and has not reviewed any financial information, he or she is assuming that the prospective earnings will provide a fair return on the assets at this level of value. The appraiser needs to say something to the effect that it has been assumed that there are adequate earnings to support the level of value concluded. If financial information has been reviewed, and the conclusions are supported by the analysis, the appraiser should say something to the effect that the prospective earnings are adequate to support the level of value concluded. By doing this the appraiser adds creditability to the work as well as providing a better service to the client.

Review of the Three Approaches

Throughout this chapter we have illustrated the basic usage of the cost and market approaches to value to machinery and equipment and briefly discussed the income approach to value a business enterprise. In our discussions we have alluded to some of the strengths and weaknesses of each approach. Many readers are comfortable with one approach and tend to favor that approach. In reality no single approach is better or worse than any other approach; it is the facts and circumstances of each appraisal assignment that makes one approach better than the others. Furthermore, the ability to recognize those facts and circumstances and to use sound judgment in applying each approach to value is the key to becoming a good appraiser.

The Cost Approach

The primary strengths of the cost approach are:

1. Its use on special purpose or newer asset properties
2. Its use for asset identification
3. Isolation of specific elements of depreciation
4. Basis for allocating functional or economic penalties

The cost approach is particularly useful on new or special purpose properties. In the case of new machines, total depreciation is usually relatively small, tending to minimize the error in estimating depreciation. In the case of special purpose properties which are not frequently exchanged in the market or do not generate revenue by themselves, the cost approach provides a logical procedure to arrive at an opinion of value. Another advantage is that its use usually provides specific asset identification; i.e., the appraiser knows the specific inclusions and exclusions for the appraisal. Its use also allows the appraiser to isolate the various elements of depreciation, which may be factors in such things as insurance settlements or property tax appraisals. Finally, the cost approach can provide a basis for allocating penalties, specifically economic obsolescence.

The primary weaknesses of the cost approach are:

1. Its inability to measure the full amount of economic obsolescence
2. The subjective nature of estimating depreciation
3. Its often being very detailed and time-consuming

The major weakness of the cost approach is its inability to fully measure all aspects of economic obsolescence. Another equally important weakness is the subjective nature of physical deterioration, which we discussed earlier. Finally, use of the cost approach can be rather detailed, especially when estimating replacement or reproduction cost. This detail can obviously be very time-consuming and expensive.

The Market Approach

This discussion will be limited to market approach as applied to individual items of machinery and equipment. The primary strengths of the market approach are:

1. Most reliable indicator of the marketplace for individual items
2. Direct measure of depreciation for individual items of machinery and equipment

The market approach is particularly useful on properties that have established markets such as automobiles, construction equipment, aircraft, and certain machines. When appraising machinery and equipment and trying to measure fair market value-in-place (emphasis on market), the appraiser should first look to the marketplace (emphasis on market), seeking data to subsequently estimate value.

The primary weaknesses of the market approach are:

1. Lack of comparable sales
2. Subjectivity of comparison
3. Lack of knowledge regarding sales
4. Timeliness of data

When there is a limited market, indicated by the lack of sales, it should be obvious that it will be difficult to measure value. Even when there are sufficient sales, we often lack sufficient knowledge regarding those sales. Such things as the motivation for buyers and sellers, financing, and condition of a machine, which obviously have an effect on value, are unknown. Using a sale which occurred during poor economic times probably is not valid for the appraisal of similar assets in good economic times.

In the discussion of the market approach earlier in this chapter, we talked about the use of the market approach as it applies to valuing an entire facility. We concluded that to do this was extremely difficult because of a lack of comparable sales.

The Income Approach

The income approach is usually not applied to appraise individual items of machinery and equipment but may have validity when the assets being appraised are part of a business. The following discussion pertains to the use of the income approach as applied to the valuation of a business.

The strengths of the income approach are:

1. Best measurement of total depreciation of all assets
2. Recognition of economics
3. Reflection of the logic and rational used in virtually all business decisions

When we conclude value for a business by the income approach, we have the value of all assets in aggregate. By doing so we are accounting for all elements of depreciation. Implicit in the development of the income approach is its ability to recognize the full amount of economic obsolescence, not achievable through the use of the market or cost approach. Finally, virtually all business decisions are made on the basis of either making money or saving money. On this basis, then, the income approach reflects the logic behind most business decisions.

The weaknesses of the income approach are:

1. Cannot segregate specific assets
2. Subjectivity of income projections and rates of return

After completing an income approach an appraiser has one value which applies to the entire property. It is virtually impossible to identify the value of specific assets without some type of allocation. In addition, it can be difficult to determine exactly what specific assets are included or excluded, especially on very large businesses. The major weakness of the income approach relates to the projections and assumptions that are made. Certainly the projections of expected future income require the use of a crystal ball, and the development of the appropriate rates of return (capitalization, discount, etc.) is subjective; both of these in combination can have a significant effect on the final conclusion of value.

Conclusion

It is quite common for machinery and equipment appraisers to discuss the pros and cons of each approach as it applies to the appraisal of machinery and equipment. Some appraisers have a bias toward the cost approach, while others are biased toward the market approach. Some machinery and equipment appraisers do not recognize the use of the income approach to value either individual assets or an entire business. Who is right? We hope we have answered the question, having illustrated the use of the three approaches as they apply to machinery and equipment. All three approaches have specific strengths and weaknesses, and the validity of each approach is a function of the facts and circumstances surrounding the appraisal. An appraiser who feels that there is only one way to appraise a certain type of property ought to look at that approach because, as we have seen, it is not without fault. We have seen that the appraisal of machinery and equipment is not an exact science, but we do have three approaches available to use and we ought to be using all three whenever practical. Knowing the three approaches and their strengths and weaknesses will allow the machinery and equipment appraiser to be more objective, adding credibility to himself or herself as well as to the appraisal industry.

Appendix

The cost-to-capacity relationship is explained by the equation

$$\frac{\text{Cost A}}{\text{Cost B}} = \left(\frac{\text{capacity A}}{\text{capacity B}}\right)^n$$

where n = size exponent or scale factor

The equation suggests that there is an exponential relationship between cost and capacity. This relationship originally was known as the six-tenths factor which "... implies that its value is 0.6. ... C.H. Chilton was the first to publish comprehensive information and data on the six-tenths factor. He provided data on some 35 complete plants which indicated that the cost curves for process plants were straight lines on log paper. Their slopes ranged from 0.33 to 1.02, but the bulk of them were closer to 0.6 and their overall average was close to 0.6" (extracted from *Cost Engineers Notebook,* American Association of Cost Engineers, Morgantown, West Virginia, June 1978, by permission).

The equation is commonly used by cost engineers when three of the four variables are known. Most commonly, it is used when a specific cost and capacity are known for an asset and the cost estimator is trying to estimate the cost of an individual asset but for a different size.

For further information on the subject, refer to the following books: *Project and Cost Engineers Handbook,* 2d ed., American Association of Cost Engineers, 1984, and Frederic C. Jelen and James H. Black, *Cost and Optimization Engineering,* 2d ed., McGraw-Hill, New York, 1978.

The equation used to determine economic obsolescence using the cost-to-capacity relationship is derived with the following steps, using two equations and two unknowns.

Begin with the original equation:

$$\frac{\text{Cost A}}{\text{Cost B}} = \left(\frac{\text{capacity A}}{\text{capacity B}}\right)^n \quad (1)$$

When the capacities are equal, the costs are equal and the equation can be expressed as

$$\text{Cost B} = (1)(\text{cost A}) \quad (2)$$

Any variation in costs (v) can be expressed as

$$\text{Cost B} = (1 - v)(\text{cost A}) \quad (3)$$

or

$$\text{Cost B} = \text{cost A} - (\%v)(\text{cost A}) \quad (4)$$

Therefore

$$\text{Cost A} = \frac{\text{cost B}}{1 - v} \quad (5)$$

Substituting equation (5) into (1) results in

$$\frac{\text{Cost B}/(1 - v)}{\text{Cost B}} = \left(\frac{\text{capacity A}}{\text{capacity B}}\right)^n \qquad (6)$$

Solving the equation for v and multiplying by 100 results in the percentage variation, $\%v$.

$$\%v = \left[1 - \left(\frac{\text{capacity B}}{\text{capacity A}}\right)^n\right] \times 100 \qquad (7)$$

The concept applied to economic obsolescence suggests that because of economic reasons, a portion of the investment is idle and provides no utility. The equation does nothing more than convert the idle capacity to a depreciation percentage to be used in the cost approach.

9
Liquidation Value Concepts

Leslie H. Miles, Jr., ASA
CEO, MB Valuation Services, Inc., Dallas, Texas

This chapter should help the reader understand liquidation concepts and their common applications and how to arrive at a conclusion. It is not meant to teach how or when to liquidate assets. Any liquidation value concept presupposes a certain set of conditions and approximates the probable outcome if similar conditions exist as did at the time of inspection and/or research. There are three commonly used concepts of liquidation which are discussed: liquidation value, orderly liquidation value, and liquidation value-in-place.

Concept Uses

The liquidation concept is used primarily for three purposes. The most common applications are seen within the collateral-based lending industry, which encompasses such terms as *secured lending, commercial bankers,* and *intermediate term lenders*. Secured lenders usually require a form of liquidation concept applied to the collateral, which establishes the basis for securing a loan. The lender wishes to know the potential collateral exposure in the event of company failure. The concept of liquidation value should be chosen by the client rather than recommended by an appraiser with the inherent exposure to values. The lender should believe that, in the event of future tests, the set of circumstances attached to the concept of value will exist to maintain credibility of the resulting indicator. A lender may fund "hard cash" using an appraisal of collateral to determine a comfortable exposure level. In addition, the appraisal listing may be used to formulate a security agreement. The results of the written report may be compared with prices established in a future sale after default. The appraiser's reputation can be on the line, and therefore the appraiser should be reasonably correct using client education techniques within the study for items of volatility. A lender can be the best referral of new business, so a reputation is best built and maintained by delivering an honest, accurate, and well-written report.

Equipment or plant purchasing is another use for a liquidation study. An entrepreneurial client may wish to know of potential downside risks prior to making an acquisition. This buyer-client may use the study to measure the collateral value (as might be required by a potential secured lender) prior to expending additional costs (e.g., audit fees). Auctioneers sometimes use appraisers respected for accuracy under this value concept prior to making guarantees or for acquiring a business to auction or liquidate. In some instances, business owners may employ this type of study to approximate refinancing potential or possibly to indicate probable results if disposal is elected under a liquidation concept.

Liquidation studies are commonly used for judicial determination. The concept is used in U.S. bankruptcy courts as a measure against a pending sale through Chapter 7 or Chapter 11, or for funding a plan of arrangement. The courts may use the liquidation concept to make a determination in allowing or disallowing the debtor to continue operating, with secured or unsecured assets, in response to other parties having a vested interest in the property. The study is often used in determining if the "automatic stay" imposed against secured parties should be lifted to allow those assets to be converted from an illiquid to a liquid state. Liquidation concepts are not always accepted in other types of courts, although there are exceptions based upon circumstances.

Understanding Definitions

The word *liquidation* is defined by Webster's dictionary as "to determine liabilities and apportion assets toward discharging indebtedness" or "to convert (assets) into cash." In the appraisal profession many terms are specifically used for a unique application, making each a "word of art." In each of the three liquidation concepts the term *forced sale* appears. The words *forced sale* tend to be forgotten in both the appraisal application as well as in the reader's understanding. For this and other reasons this text was written using formally defined terms as standards for the industry. These formal definitions are meant to prevent confusion which may arise when *forced sale* is used as an appraisal term.

Liquidation can be accomplished with or without a forced sale application. However, any disposal without using some form of forced sale application would be guided by the fair market value concept. Such liquidation may set a precedent establishing another fair market value. For example, we may look at comparable machine tools as priced by various dealers within the same region. Almost identical Bridgeport mills from three different dealers within the same geographic area may have prices which vary by hundreds of dollars (e.g., $4800, $5100, and $5300). Which of the three values should be considered to represent the retail, or *fair market value*? Each sale may be made with some standard guarantee, which could then establish a retail fair market value or that amount paid to a dealer. If those same dealers, not forced to sell, were to sell these machines on an "as is/where is" basis, the amount recovered might be less and thereby would set a comparable fair market value without warranty.

In any event, the dealer is assumed to be selling these machines with no forced requirement to do so; this would be true if the dealer wished to sell one machine or to dispose of all equipment anticipating a closeout of business. However, a dealer who prescribes a time limit within which all equipment must be removed from the premises creates, in effect, a *quasi forced sale* condition. A better explanation of *forced sale* could be that the current owner will lose control in the disposition of the assets at some predetermined point in time. This loss of control could be through bankruptcy, foreclosure, voluntary turnover to a liquidator without reserve, or other circumstances in which the owner would relinquish rights to the property owing to outside influences.

By utilizing standard defined terms ("words of art") an appraiser can communicate with clients as well as peers. If an appraiser were to state, "I am doing a liquidation value appraisal," it should be understood that the resultant value is consistent with that which would be derived from a properly advertised and conducted public auction sale held under forced sale conditions. However, an appraisal accomplished for liquidating a client's business should use a Fair Market Value concept. To liquidate is to dispose of as defined by Webster, whereas to conduct a liquidation value appraisal should be to predict the outcome of an auction held under forced sale conditions.

The purchaser of any liquidation concept study should be made aware of variables which affect the use of any itemized valuation estimate. The study should not be used to sell from but rather be used as an internal guide for what to accept. Asking prices or individual sales should be made from a fair market value guideline. Liquidation studies are average value estimates adjusted for various causes and effects. In application, the actual sales may vary from the appraised indicators with the resultant total in general agreement with the appraisal. *Averaging* is the method for arriving at an overall conclusion. This understanding is especially important for orderly liquidation but can also be a consideration for the other concepts. Individual sales should always be made from a fair market value guide, and the results will, in all probability, be the liquidation study total indicator if the conditions associated with the concepts are equal.

To understand the liquidation appraisal concepts in theory as well as application, we expound on their definitions in the following sections.

Liquidation Value

The definition of *liquidation value* takes into consideration such inflationary or depreciable conditions as physical location, difficulty of removal, adaptability or specialization, marketability, physical condition or overall appearance, and total psychological appeal. The concept evaluates the potential of assets to draw interested buyers. The assumed auction sale indicator is a proposed gross sale without prior knowledge of any unknown expenses such as auctioneer fees, setup, marshalling, court costs, or advertising expenditures. It assumes that all equipment would be sold on a piecemeal basis, *as is* and *where is*, with buyers being responsible for removal of purchases at their own risk and expense.

The indicator typically cannot assume additional values that may be derived from such an auction sale, including but not limited to product line, equipment-in-place, going concern, patents or recipes, rights to manufacture, trademarks, mailing and/or customer lists, and jigs and fixtures. Even though the auction sale could possibly produce bidding for these items, such higher return cannot realistically be anticipated by an appraiser.

Orderly Liquidation Value

This concept differs from an auction sale (liquidation value) because of the extension of time allowed for piecemeal negotiable sales. Forced sale conditions exist in that all items must be disposed of at some specific future time. The time associated with orderly liquidation value depends on the scope of the project; some appraisers may use 60 or 90 days, six months, or even a year. This concept carries with it the assumption that a knowledgeable liquidator would be able to properly advertise and merchandise the listed assets in order to stimulate a correct response, which creates the consummated sales. This value must, again, take into consideration physical location, difficulty of removal, adaptability, specialization, marketability, physical condition, overall appearance, and overall psychological appeal.

In most cases, it assumes buyers would be responsible for merchandise removal at their own risk and expense; some exceptions require the liquidator to cover shipping and handling. An alteration to the time of disposal can cause a dramatic change for anticipated dollar recovery. This time requirement may be dictated by either the appraiser or the client. A client's specific time limitation can alter the appraiser's value indication from a standard time that would better fit the subject assets.

Liquidation Value-in-Place

This concept is the most misunderstood of the three under discussion. It is one not often applied owing to the higher potential risk associated with obtaining the indicated value. It assumes that a failure or nonprofitability of the business is such that the plant and equipment must be sold. It considers that a purchaser would be available with the belief that a new management team or director would make the company profitable. It additionally assumes that the reason for the failure or lack of profits was not due to the current economics of the industry, the product manufacturing flow, or the equipments' use.

In most instances, the assumption would be that management was the cause of failure or that past economic conditions (such as recession) degraded the company to a point requiring a sale despite a promising future. The majority of plants that fit the liquidation-in-place concept include tank farms, chemical plants, steel mills, and other industries which have high costs for special installation and therefore limited adaptability or marketability on a removal basis. It is often used by lenders when it is the only forced sale concept which could support a collateral-based loan. In some cases, a removal concept would indicate a negligible value insufficient to support the typical collateral-based loan.

This concept encompasses industry economics and the plant location which takes advantage of the economics. There should be an investigation of the need for such a facility in its present location before relying on this concept. The appraiser assumes that the land and buildings can be used as inspected, that fair and equitable lease arrangements would be acceptable to a purchaser, and/or that the real estate and improvements would be a part of the hypothetical acquisition. Where research reveals that no buyer could be found who would fit the concept assumption, the appraiser should advise the client to alter the study parameters. However, where the concept is requested and is potentially possible without regard to the degree of probability, the value could be used. In any case, the client should investigate probability and inherent exposure of such derived value.

The reader is warned to be very careful in the application of the liquidation-in-place concept without sufficient exposure and experience in this area. This section of the chapter is not meant to teach this concept application and merely generalizes the theory as an overview. It is, more than any other, a concept subject to tremendous swings in value opinion by those of different and possibly limited experiences. More importantly, the users of this type of study may improperly reflect on its result and make an improper monetary judgment, which could affect the reputation of the appraiser. It is a concept that requires experience in proper relationships, in client education and acceptance with full knowledge of its special meanings and volatility, in the ability to know when to reject the assignment when it may be used improperly, and in the knowledge of the potential liabilities that may be reflected back to the appraiser and affect his or her reputation through improper comparisons on a later analysis. Used properly in application, judgment, and theory, liquidation-value-in-place can be a valid and reasonable concept.

Experience Required for Liquidation Value Concepts

When applying the forced sale concept the appraiser should be knowledgeable in three specific areas relating to that value:

1. Values as altered by the set of circumstances associated with the particular concept

2. Industry nomenclature applied to the specific types of equipment to give a proper and full description that identifies each item (this includes the ability to know when items of equipment should be described as a unit or valued piecemeal)

3. Economics of identical or related industries in order to make a judgment that may alter a final value assignment applied by the appraiser (there are many causes and effects which alter value, and the understanding of these can be gained only by personal experience in the marketplace; general training serves to supplement a thorough understanding gained through experience)

Auction Sales

Auction monitoring contributes most to understanding this value concept and the related causes and effects. On-site monitoring is not only essential in the training phase but also should be ongoing throughout the appraiser's career. The sale monitoring is beneficial in two ways for the appraiser:

1. It allows the appraiser to observe what can affect a sale's outcome, such as weather; psychological appeal, which may be associated with the improvement that houses the equipment; location; difficulty of removal; total appeal of the equipment as a group, which may attract a proper attendance; and the way equipment should be appraised as would be anticipated if it were sold. This observation of causes and effects aids in value orientation.

2. The gathering of comparables can be kept for future references in appraisal application. Other ways to gather comparables are to trade information with others or have sales monitored for a fee for which the information is supplied later. If enough outside information is gathered by others, relationships can be drawn between similar items of equipment to establish data accuracy.

Equipment value is best estimated by considering sales of equipment in the same industry or, although less desirable, in a sufficiently similar industry, which would allow adjustments to be made by an experienced appraiser. When there is no alternative, a judgment is made about a relationship to another industry where stronger comparables can be found. The appraiser may estimate value by a reference to sales in an industry with similar economics.

Sales monitored by outsiders can have mistakes through misunderstandings, errors in posting, or simply someone (seller, buyer, or agents) not wishing to give the proper requested information to anyone other than potential buyers. Monitored sales, even by outside sources, can provide excellent mass information for use in establishing value relationships. An experienced appraiser, understanding liquidation sales, can interpret monitored information and derive reasonable value estimates. This allows the appraiser to question supplied comparable data which seem to be inaccurate or abnormal.

Understanding the Industry

An appraiser within a particular industry must become familiar with it, if only in a general way. Familiarity will allow the appraiser to discuss the job with a client properly so as to gain additional information in areas that could require further explanations or investigation. An appraiser who can relate to only one type of industry is limited in drawing relationships to that which he or she has seen sold or appraised; i.e., contractors' equipment cannot be related to a specialized industry such as chemicals or food processing. In some cases, it might be possible to have enough experience with various types of equipment that a relationship can be drawn to other manufacturing areas.

If the appraiser has experience with contractors' equipment, metal and woodworking equipment, chemical plants, food processing, textiles, garment plants, etc., there may be a substantial amount of this equipment in another industry that allows the appraiser to make proper value relationships. The more varied the experience, the more diversified the appraiser. It is necessary to observe which types of equipment will bring higher return on cost and which types are difficult to sell. An auction sale can offer insight into marketability and can provide a yardstick for use in the other liquidation value concepts.

Finally, it is necessary to observe special conditions which can cause a higher or lower than normal return. Additionally, the appraiser should recognize when special conditions do not exist. An example of a special condition could be the appraisal of a boiler. We assume that the boiler had a value in liquidation of $30,000 removed and a value of approximately $37,500 installed. If it were placed in a room with a wall that did not allow access adequate for removal, the buyer under a liquidation value removed concept may adjust bidding at or above an anticipated cost of removing the wall and repairing it. It is possible that this wall may not allow a liquidation value to be placed on the boiler.

Another example is to have specially built equipment for which exact comparables may not be available and relationships must be made. It may be that most, if not all, companies within this industry manufacture the same piece and an acceptance would be standard even though shop-built. One other example could be an engine lathe with special adaptations that may or may not affect return at the liquidation sale. If we assume that special adaptations would not be acceptable as installed, we must determine if buyers would readily accept the machine as a standard lathe on which the adaptations could be removed easily. There are times when adaptations cause an increase and times when they cause a decrease, even though affected only by a psychological impact. There are also times when special adaptations cause no alteration in a standard machine's return. All these special conditions must be observed in the liquidation marketplace as observed by the appraiser or appraiser's aides. Various examples can be found in Chapter 2, "Classification of Property."

Communication

It is necessary to develop and maintain a line of communication with dealers, liquidators, and other appraisers. Such communication provides information necessary to make adjustments to a particular study and better understand comparable sales. Individuals within specific industries can provide information on relevant economics at the time of a study. Lines of communication with others provide information that can help with terminology and use of special terms and may aid in future studies. Dated comparables help plot historical trends which, when necessary, can be used for residual forecasting or indexing to a study date.

The most important area of required comprehension is that of obsolescence. Communication coupled with experience will allow recognition of current obsolescence as well as anticipation of potential future obsolescence. Industrywide knowledge of an item's obsolescence can affect the current value return; computers provide a prime example. When current comparables are unavailable, relationships and opinions derived through communication contribute greatly to assigning reasonable values.

Methods of Valuation Used for Liquidation Concepts

More so than any other concept, a liquidation study should be listed by an experienced appraiser who has personally seen the subject. It is impossible to understand all the combined causes and effects using only an equipment list. This does not mean that others cannot help produce a listing, but the final assignment of values should be made by an experienced appraiser who has observed the group of assets under appraisement. This type of appraisal requires a high degree of accuracy for two primary reasons:

1. This type of study is normally used for loans, and a lending institution may advance money based on the report.

2. The appraiser may jeopardize his or her reputation if the study is deemed unreasonable due to erroneous assumptions.

The Listing

The appraiser should list all items under appraisement and show individual average pricing. The description should properly identify the items and, where possible, record condition notes. For later research, the on-site listing should carry with it the general field and condition notes so that they can be used in later correlation. It is usually very difficult to remember all details upon returning to the office. On-site notes, referred to above, are over and above text which would be made a part of the published report.

Research Library

After the on-site work has been completed, the key element of an appraisal study is the research, guided by the particular value concept(s). However, as this chapter primarily covers liquidation concepts, it does not speak to utilizing this type of research toward other concepts of value although it may generally apply.

Establishing a good research library is crucial to an appraisal firm wishing to establish credibility utilizing liquidation concepts. The library should be set up in such a manner that comparable information is readily accessible. Libraries may be organized with manual or computer-aided systems and contain standard subscription manuals containing relevant information. In addition, a library should contain trade journals, tabloids, magazines, newspapers, buyers' guides, manufacturers' brochures, and additional publications related to the specific industry or industries in which the appraiser or appraisal firm will take assignments.

If liquidation comparables are not available for some industries, relationships will have to be made for projecting a final judgment as applied by a qualified value expert. Judgment is generally tempered by other types of information gathered by outside sources. It is necessary to know how to locate those sources, and the library can contain that type of information. It may be necessary to call a manufacturer to obtain information about a new cost, the history of any resale of a manufacturer's equipment and/or its acceptance in the used marketplace, technological obsolescence, general demand through back orders or their lack, and the quantity currently in the marketplace. Good contacts should be established to aid in determining the used market acceptance, asking prices, and used prices paid.

Resale is best determined using recent comparables of identical equipment. A sufficient number of current sales may indicate the average condition of a market. Comparables are usually not found on all equipment, but those that are gathered should be kept in an acceptable and retrievable format. Telephone survey information can be kept for future use. It may be helpful to use standard comparable data forms to record and file information in the library.

Monitoring auction sales, discussed earlier, is a prime source for obtaining liquidation value comparables and can be helpful in establishing trends or general marketability. Bulk sales or negotiated sales after failure for an entire business may support values and adjustments for liquidation value-in-place. Most judicial sales are public records, and it is wise to obtain and keep this type of information. As all manufacturing facilities have differences, bulk sales with their variables may not have direct application to the subject. The knowledge of these sales, however, can lead a value expert to form a reasonable conclusion as adjusted for a particular set of circumstances for a given group of assets.

The hardest comparable information to gather is for orderly liquidation value. Used prices are the best guide to one experienced in orderly liquidation. However, some orderly liquidation sale information can be discovered through banks or commercial lenders who have had to perform this type of asset dissolution. Many lending institutions will supply this information under a reciprocal agreement of the appraiser since the data bank could be helpful to them in the future. In some instances, dealers or liquidators could be helpful in supplying sales information if they understand that information sharing is reciprocal.

Correlation

After collecting data, the appraiser must draw relationships for the concept being applied. When comparables are unavailable, prices for new, like, or similar equipment in the marketplace may prove useful. It is possible that a comparable known sale of another item could be used in comparison to its new price and then applied to the subject. A Bridgeport mill could bring $3500 and then, compared to its new cost today, be related to an index mill of like capacity and condition. Naturally, adjustments would be made according to general acceptance of the index mill in the marketplace. However, it is possible even to use other types of equipment that are unlike the subject and still make the same relationship.

There could be equipment in one industry that has similar characteristics to a subject located in another. The appraiser would make adjustments but would be convinced that there were similar economics and/or measurable demand characteristics for the comparative item. Also, used prices of other types of equipment having similar demand or economic characteristics could be applied and adjusted to the liquidation concept being applied. A used price of an item could also be related to a known standard fair market value and/or new cost and then adjusted to the subject. A comparable liquidation outside the industry can be compared with either of these two known factors (fair market value or new price) and then adjusted to the subject. On an itemized basis, the above could hold true for all liquidation concepts if adjusted by the set of circumstances that would follow the specific definition of the value.

An adjustment must be made to what is labeled *causes and effects*. Four main factors make up that terminology: total draw of the items, physical appearance of the items, industry location and economics, and psychological effects.

Total Draw of the Items. An item in one plant may not have the same liquidation value as one in another plant even though the plants may be side by side. Part of the reason for this may be that relationship of *total draw*. An older forklift may bring less at an auction sale in which there is only old and unattractive equipment and/or very little equipment making up that sale compared with that same forklift being in a million dollar sale with very attractive equipment. This means that the overall appeal of the assets influences the individual pricing that is done within the listing.

By the same token, if part of these assets were removed, it could affect the item's average price that the appraiser had originally applied. Appraisers should be very careful to qualify this possibility so that future tests with less equipment do not cause an improper comparison that would leave an impression that the appraiser was originally in error.

Physical Appearance of the Items. Liquidation sales, for the most part, are conducted over a shorter period of time than other approaches to value. Auction sale recovery can be highly dependent on the appearance of equipment, more so than actual condition. Naturally, the known condition can affect residuals, but auction history indicates dramatic effects through appearance—both higher and lower. The appraiser should not simply take the appearance of the equipment as inspected but rather as the appearance would be if set up for a properly advertised and conducted auction sale. Such presale preparation includes cleaning and, possibly, nominal repair.

The appearance of condition should be a primary consideration for liquidation value application. This appearance of condition is also important for orderly liquidation value, although tempered somewhat by demonstrated condition due to the time element allowed for closer inspection by a prospective buyer. Cosmetic appeal is important to the liquidation-in-place concept but, further, to the operable and functioning portion of the plant. Equipment not in operation must still be discounted under liquidation-in-place if it is known that the equipment does not contribute to the current operation for which the concept is totally reliant.

Industry Location and Economics. Certain industries are made up of equipment relating only to a particular type of production. That type of equipment can be judged as having a higher return during the "good times" of that industry and lower during the doldrums. This would not necessarily relate to equipment, which has versatility within many different industries, all of which are not depressed. In some cases, a plant location can affect anticipated return in liquidation. If it is anticipated that the majority of equipment would be sold to remain in a particular area, the study may indicate a value far less in that area than in another. This can be especially true for liquidation-in-place, where the outcome, owing to economic conditions in a particular location, might be affected.

Psychological Effects. Psychological cause and effect is much higher with regard to auction sales (liquidation value) than to the other two liquidation concepts. It is possible that equipment located in an older, run-down building would bring less at auction than the same equipment located in a modern facility. The reason for this is considered one of psychological impact, e.g., an executive desk of good quality being sold in a beautiful executive suite as opposed to that same desk being offered in a warehouse surrounded by old machines.

There is a contributing factor to the surrounding or placement of items offered for sale. Orderly liquidation can have impact on surroundings regardless of the sales craft that goes into the negotiations. However, some psychological impact may be overcome by sales ability. Liquidation-in-place is the least affected by psychological effect other than the impact of failure. Failure may have some impact on the former customer base and can attach itself to the equipment and its location even after a name change or new corporate structure. This then would relate to psychological impact from outside forces as applied to the equipment.

The primary guides for value indicators are used prices, paid or asked, in the resale market. Naturally, the main emphasis is on historical prices that are paid for like items. However, if enough asking prices are available in the marketplace, an opinion can be derived for the fair residual value of the subject. The previously mentioned method of research related to obtaining used asking prices in the marketplace. It is also necessary to obtain prices paid for used or new equipment other than those from monitoring auction sales or contacted sources.

An excellent additional method for obtaining comparables would be through client documents such as depreciation schedules which show cost and date of acquisition, invoices of major purchases, asset ledgers, etc. A request for this information is usually answered affirmatively and becomes an additional and viable resource or information-gathering tool. With a little extra effort, this information could also be placed into the appraiser's data bank when the item described in the appraisal can definitely be identified as the one posted in the furnished documents. Over the years, this information can become valuable in establishing trends as well as drawing relationships with the concepts being applied. This then gives prices (new or used) of equipment the appraiser observed and then described in a manner consistent with what may be done in the future. The used market establishes residuals and becomes a primary guideline for value assignment to even new equipment.

As we know, an item purchased new for use becomes used immediately and must be valued as such within the historical depreciation guidelines that follow it.

Conclusion Derived from Correlation

After research has been completed, a conclusion can become obvious if all things are considered. For an expert on a value concept with all its various causes and effects, comparables can be an excellent guideline. However, other areas must be weighed to give the appraiser a comfort level for the final indicator. These additional areas are intuitive value indicators, positive marketing aspects, negative marketing aspects, and special considerations.

Intuitive Value Indicators. In some instances there are no comparables, and judgment must be made. Experience in the marketplace can allow value orientation to diversified items because they relate to the historical return of different items. Intuitively a judgment may be required on this past value orientation as applied to some areas that would have little historical support. It is surprising how intuition can also lead to further investigation that may correct errors gained in research. Researched items can have improper posting, false information, errors in size or condition of the comparable, or special sale considerations brought out by the research. The intuitive value indicator should be 25 percent of the final analysis moving toward a final conclusion.

Positive Marketing Aspects. The next 25 percent of correlation measures the marketplace as applied to the positive points of the item and the requirement (demand) for same within its current industry or field as well as any alternative fields (secondary markets). Positive marketing aspects could be air clutch on a punch press, popular sizes, good marketplace reputation, brand recognition, and versatile market share.

Negative Marketing Aspects. It is impossible to measure positive marketing aspects without investigating the negatives. The negative and positive marketing aspects should be combined as 50 percent of the correlated prices. Examples of negative marketing aspects are such things as specific industry applications in which there are poor industry economics, a lack of Occupational Safety and Hazard Administration required protection, declining acceptance due to age or newer technology, or lack of support from the manufacturer.

Special Considerations. The final 25 percent is a look at other areas that may be forgotten. These considerations could be positive or negative marketing aspects, but can be overlooked. It is typical to look at conditions, general marketplace, general industry economics, etc. Sometimes overlooked is the area of adaptability through possible alternative uses. Many years ago the standard planers (metalworking) were converted to profiles or planer mills which caused the basic frame of the planer to be more saleable. Even though things have changed since then, this bit of history shows that equipment may have alternative uses that should be considered.

By the same analogy, alternative uses may be considered and found to be minimal at best and not necessarily creative of a measurable alternative market. Another variation of alternative use that the appraiser should recognize is the ease of reconverting an item to its standard configuration. A low number may be applied to a special adapted forklift if the appraiser does not realize that it could quickly be returned to its former configuration, which would be recognized by a potential purchaser. In some cases, an alteration back to standard could be done prior to offering it for sale, which would then be assumed under any removal concept. To reiterate, alternative use could be an adaptation of a less marketable piece of equipment to obtain a greater market share or to convert from special use back to a standard that has more acceptance. Experience coupled with judgment are the tools that allow the appraiser to make these measurements as to the reality of that happening in arriving at a final value conclusion.

A second consideration is applied to the removal concepts of liquidation; more specifically this would be used under the definitions *liquidation value* and *orderly liquidation value*. A buyer's consideration that would be applied to the difficulty of removal must be understood. A boiler made of brick and mortar may have no removal value owing to its inability to be reused for the purpose intended after removal. A boiler in a room which has been built around it may not have the ease of accessibility for removal even if it is a standard package type typically sold on the used market. In utilizing the removal concept, it is absolutely imperative that the ease of removing that item be anticipated by the appraiser as would be applied by a potential purchaser. Sometimes equipment that is mounted in pits or concrete can be removed but discounted for historical happenings as observed by the appraiser. It could be said that the appraiser is measuring difficulty or ease of removal in the application of the final indicator.

The above considerations make up 100 percent of what we refer to as the *final analysis for application to comparable indicators*. They take into consideration all the causes and effects for the final required adjustments. Experience will bring an ability to perform the correlation, in all its aspects, quickly and efficiently. This then would leave the time for significant items for which there is more time required. The appraiser's report should go out only when the appraiser is certain that the indicators are proper and fair under the concept being applied.

Report Assembly

In a standard appraisal the cover work is straightforward and understood by most readers in the profession. It is reviewed here to bring out certain specific points as they may relate to the liquidation value concepts. At the beginning there should be the *letter of transmittal* indicating to whom the report is addressed, the general understanding with regard to the scope of the project, and any specific references that need to be made for there to be no confusion with the content. There should be a *recapitulation page* indicating the overall value for the concept application as formulated through the itemized listing. To avoid confusion, the recapitulation page should restate the concept identifier.

The *definition* should be made a part of the report and, if applicable, should use concept titles as shown in this text. The section for *statement of limiting conditions* should contain any standard and/or unusual areas that need qualification. If all vehicles were not seen, it should be stated at this point. It is imperative that the reader be able to understand totally the report and any assumptions that were required in producing it. This will allow the reader to make a good judgment for monetary considerations, which is the report's typical use. Although not required, the *method* could be stated in a general way that will give the reader some confidence in the writer as well as an understanding of an approach to value over and above the reading of the definition.

If there is a requirement for *special considerations* under a liquidation concept, it is advisable to make that section a separate area of the report so there is no confusion. A request may be for the appraiser to judge the value of dies or molds which, although intangible, are typically considered having an intangible value under a liquidation removed concept. If this is highlighted by, and placed in, a special considerations section and further highlighted in the *recapitulation* without being included in the total, there will be less chance of a wrong comparison of the appraiser's study to a future test. Special considerations are simply those which go beyond the general content of a standard liquidation-type study. If the study is computer-generated, it may be necessary to include an *interpretation page* to ease the reader through the report. Interpretation pages are good for cross referencing one section of a report to another when that is required.

Once the cover work is in place, the following data (itemized listing) can be reasonably understood as being the opinion of the appraiser qualified by all that precedes it. If the *certificate* is to be applied (certification of appraiser), it typically follows the data and is associated with the *qualifications* of the appraiser.

Nonstandard Studies

There are other printed studies that are not necessarily appraisals. The word *appraisal* is synonymous with dollar value interpretation of an item. However, it seems to be associated with an absolute quantity and can affect the reputation of an appraiser. Because of this interpretation, the word *appraisal* should not be used in nonstandard studies in which either proper research and/or on-site observation has not been accomplished. There are four types of value studies that should not be considered appraisals but rather *value opinions*. It is best to leave the word *appraisal* out of these types of studies other than to make a statement that "the value opinion is not an appraisal." These types of studies are merely guidelines and may be subject to a later appraisal in order to arrive at a final and correct number. The following studies are what should be considered value opinions: walk-through studies, desk-top (not seen) studies, hypothetical situations, and intangibles.

Walk-Through Studies

Under a liquidation-type concept, walk-throughs have become almost commonplace for giving quick value judgments. This type of study allows the one giving the opinion to at least observe the items for a better understanding of the causes and effects. Many individuals make this on-site walk-through quickly while making mathematical computations. There is usually little or no research. The value judgment can be generated quickly when time is of the essence. There are, however, some major pitfalls with this type of study. The obvious problem is one of the appraiser's accuracy and possible comfort level with the indicated value. Also, the user of that value indication will, in all probability, consider the number or value range accurate regardless of how the appraiser qualifies it. The reader will expect a later appraisal, properly conducted, to be at or close to that value indicated in the walk-through.

Because of the user's probable interpretation, it is up to the valuer to consider this type of exposure in the acceptance of the assignment without having the standard remuneration for same. A walk-through is simply a cheaper process of arriving at a value and should not be used with the same reliance which an appraisal, with its standard fee, allows. A walk-through report should contain an itemized list even if derived from a cursory look. Itemized values need not be shown, but an attachment of the listing to an overall opinion can avoid later misunderstanding.

The list indicates to the reader what was observed in the event items were overlooked and/or included that should not have been. Above the reader's use of this listing, the valuer should be aware of a potential future test through a comparison to a later conducted appraisal or sale. It could be that the future test would not include all of or the same items inspected on the walk-through.

The only way to confirm this is by a listing previously made on the original walk-through study. It is essential that a walk-through study contain some sort of listing. In addition, it might be wise to always include in the walk-through study a range of values rather than an exact number to allow latitude for a potential later comparison to a more thoroughly researched appraisal study.

Desk-Top Studies (Not Seen)

A *desk-top* study contains an opinion of value without physical inspection. It requires assumptions that then qualify the study. Experience has shown that desk-top assumptions can be highly unreliable when compared with an appraisal after an on-site study has been completed. Some appraisers avoid this type of opinion for the same reasons expressed in the section on the walk-through study regarding acceptance by the users of a value opinion. Experience has shown a better acceptance and understanding by the clients of desk-top studies regarding value alterations after a later on-site inspection has been completed. However, many use the desk-top opinions to make monetary decisions which are, in reality, credit decisions.

The proper use of desk-top studies is to measure the possibility of an outcome if a future formal study is later conducted. Other uses are to forecast residual value, to offer opinions of general industry economics in which the equipment may fit, to allow an expression of views with regard to the marketability of the type of equipment, and in some cases to measure the price paid against the typical market. There are other uses of desk-top studies which again should be qualified as opinions and not appraisals.

Hypothetical Situations

There are times in which qualifications within a study may require the word *appraisal* to be avoided. If a report were to qualify that the equipment being valued assumes the complete repair and/or overhaul of all items observed, the study would, in all probability, be anything but an appraisal, and rather a value forecast. The appearance of the equipment in the mind of the appraiser might contain an assumption of what the case might be if all assumed work were carried out. It is possible to see an item under repair and assume the value is as if repaired within the scope of an entire appraisal study. However, it is another thing to carry assumptions to the point that the greater part of value is dependent on future happenings.

With special emphasis on liquidation values, hypothetical situations may or may not come to pass, and the monetary decisions made could later be detrimental to the user of the report. It has become common practice to take trucking firms or plants with large quantities of vehicles and appraise them from a list provided. When a fair portion of the equipment has been observed and related to the list, the practice is considered reasonable and fair. This type of assumption sits within a gray area of being a hypothetical situation. In these types of cases, it would be unreasonable to burden the client with the exorbitant cost of seeing each item actively in use for the production of income to the subject.

This assumption procedure should be used when it is clear that the equipment not observed is in operation and that some of the operating equipment was seen in order to make a proper comparison. When equipment is not seen, it should be qualified to what the appraiser in fact did or did not see so that the reader may appreciate the report in the manner in which it was conducted. It is up to the appraiser or valuer whether a report should be called an *appraisal* or *value opinion* for the potential exposure that may take place in a future comparison test.

Intangibles

Some appraisers value intangibles when qualified to do so and where the appraiser believes that the end result is reasonable. Under liquidation values, the monetary commitments become a key consideration in how an intangible portion of the study has been conducted. If the primary emphasis of the study is in regard to individual support equipment, values are sometimes requested to be placed on each one or group. The value placed on intangible areas is usually qualified and/or conditioned on information furnished by the principal. This is not to state that intangibles cannot be appraised but rather is to make the point of how the number is applied in order to call it an appraisal.

In all probability, there should be a special section for intangibles in which there would be no question to the reader as to how that area contributes to the overall value. If the value application is merely an assumption based on information furnished, it should be qualified as such and indicated as an opinion rather than an appraisal. The value(s) of such areas should not be included in the portion of a study that is considered as an appraisal. Intangibles include such items as proprietary software (developed by and only for the user), product line (which would contain such items as special tools, dies, fixtures, patents, recipes), trade names or trademarks (good will), and customer lists.

The main thrust of this portion of the chapter is to point out the importance of potential later uses that could be applied by others. The points made are to protect the reader from making wrong assumptions but are primarily aids to the appraiser in avoiding future comparisons that could give unfair criticisms to the writer of the report. If appraisers continue to educate their clients, the qualified appraisal and/or value opinions will be better understood.

Anticipated Liability

Appraisers have an obligation to their clients, other possible readers of their reports, and themselves. They need to understand what can happen in the future with regard to their appraisals. As appraisers go through the process of dollar value assignments, they should think about future use by others and liability to themselves. Appraisers should be ready and willing to answer questions regarding justification of values they applied in a report for which they have been paid.

More than just a value opinion, an appraisal carries with it an obligation to justify some areas within reason. The application of values may be simply a relationship or judgment in which there was an attempt at a market analysis for which no data were obtained. The purchaser of the report is entitled to present his or her side for other possible overlooked justifications and/or to at least understand the appraiser's point of view within a concept of value.

Many appraisals indicate that the appraisers or writers of such a report are not required to testify. A liquidation concept carries with it the potential requirement of court testimony. If an appraiser is not willing to testify under liquidation concepts, the job assignment, in all probability, should not be undertaken. Anyone can be subpoenaed regardless of how a report is qualified regarding testimony. The appraiser should be ready to testify and be willing to support the context of the value study.

Accepting the liabilities for the potential justification of a report as well as testimony will add to the credibility and reputation of the appraiser. Liquidation value is the most provable number owing to the availability of observed sales as well as the concept of value being put to direct tests as previously defined. Fair market value concepts can have variables through negotiations that may alter a number in a future test and thereby cause no black mark on the reputation of an appraiser. However, liquidation value (auction sale) can be plainly defined as what may happen to the equipment included in a study if put "under the hammer" at its present location. Other than the economic changes, there are very few variables whose differences can be justified through a comparison of auction results to the original "appraisal" study.

If everything is equal (appraisal assumptions to actual tests), the results should be reasonably within the guidelines of the study. This holds true in all concepts of liquidation but is more easily measured under the liquidation value concept. This does not mean that the results of a sale cannot be different with justification(s), but is to indicate that the reader may make comparisons, whether proper or not, that may affect future ongoing business or relationships with those who made that comparison.

Orderly liquidation value and liquidation value-in-place are harder to test. *Orderly liquidation* presumes that all items can be sold in an orderly manner within a specific length of time. The value indicator should be reasonably correct. It is possible that all items may not be sold within that length of time, and thereby goes some liability with the production of such a report. The user of an orderly liquidation value study may consider that the appraiser is recommending that the value is the only method that should be applied to later obtain the indicated results.

In reality, the client requested orderly liquidation value with the assumption that all items would be sold to buyers in the marketplace over the period of time for which the equipment is justified to be exposed in that marketplace. The failed concept of orderly liquidation value gives a price incentive for all items to be sold as indicated. There must be individual pricing (value) under the orderly liquidation concept to do a proper study. If some items are left unsold, the appraiser may be criticized as not having indicated those items that were not liquidated in the time frame. The very idea of an orderly liquidation assumes that the liquidation sale will take place in such a manner as to maintain a proper balance so that all items may be sold. If this did not occur and/or was simply improperly liquidated, the accuracy comparison (test) may be in error.

Liquidation value-in-place is the most volatile of all liquidation concepts for making future comparisons. It assumes the purchaser is under forced sale conditions no different from the way fair market value assumes a buyer is without forced sale conditions. If the liquidation-in-place sale does not take place and items must be liquidated piecemeal, it is possible that the appraiser may be criticized for having performed an improper value concept.

There is a higher degree of liability with regard to the liquidation concepts as indicated in this chapter, and this should not be taken lightly. It may be wise to include this general statement as part of all liquidation studies:

> The value concept has been chosen by the client and should not be considered a recommendation by [the appraisal company] as to what might result in any later tests at liquidation. The concept probability and/or feasibility is beyond the scope of an appraisal. The user of the report is to determine

the probability of occurrence. The appraisal is purchased in order to allow an opinion of value under an assumed set of circumstances, as requested and mutually agreed upon by the client and (the appraisal company).

Overview

In this chapter, an attempt has been made to orient those with interest in liquidation concepts. The chapter should also aid in performing the job of standardization and peer communication. Once appraisers are all saying and understanding the same things, the better it will be for the appraisal profession. The difference between appraisers may be a value judgment, special client treatment, additional requests, or timely report production. The main thrust of this chapter is to conform the methodology in which the resultant value differences can be reasonable from one appraiser to the other. With appraisals being anything but an exact science, appraisers will differ in value judgments. There is a reasonable range at which this difference in judgments should be acceptable. An example of this would be one appraisal at $1,000,000, one appraisal at $800,000, and another at $1,200,000. It could be said that only one of these appraisals is correct, but they are all within a reasonable range that could be acceptable. Using the same analogy but different numbers, one appraisal at $1,000,000 and another at $2,400,000, performed under the same concept, is an unreasonable or unacceptable range and one of those appraisers must be wrong. With proper investigation and concept understanding, the appraisers using liquidation concepts will be more reasonable in their differences and any communication, in order to clarify, will come from a proper understanding among all parties.

10
Insurable Value

Kal Barrow, ASA
*Manager, Appraisal and Valuation Services,
Arthur Andersen & Company,
New York, New York*

Early Development

Mentions of valuations for insurance purposes have been found in the United States and England since the seventeenth century. In England, for example, Samuel Pepys (1633–1703) had to value cargoes being imported as part of his role in the Admiralty. It was during this era that Lloyd's of London came into being.

In the United States, fire protection existed in the mid-seventeenth century with the advent of fire wardens and bucket brigades. It was not until 1752, however, that The Philadelphia Contributionship for the Insurance of Houses from Loss by Fire was started by Benjamin Franklin.

However, the basic concept of value which gave way to general appraisal theory did not occur in this country until 1898. It was at this time that the United States Supreme Court rendered a decision in *Smyth* v. *Ames*.[1] Although the case dealt with railroads and the State of Nebraska, the decision stated, generally, that "...all factors or evidences of value must be considered and given such weight as may be just and right in each case, as determined by sound judgement".[2]

It is easy to see that the coupling of the needs of fire insurance poli-

---

[1]*Smyth* v. *Ames* (No. 49), *Smyth* v. *Smith* (No. 50), *Smyth* v. *Higginson* (No. 51), 169 U.S. 466, 546 (U.S. Supreme Court, March 7, 1898).

[2]A. Marston, R. Winfrey, J. C. Hempstead, *Engineering Valuation and Depreciation*, 2d ed., Iowa State University Press, Ames, Iowa, 1982, p. 20.

cies with the basics of valuation theory was a natural phenomenon. One was the ideal complement for the other.

Definitions

For insurance purposes, the following definitions are to be utilized:

Insurable replacement cost is the replacement cost new of the item after deducting the cost of the items specifically excluded in the policy, if any.

Insurable value depreciated is the value remaining after deducting depreciation, based on an analysis of age, condition, serviceable life and utility of an item, from the insurable replacement cost.

Requirements

The requirements of an appraisal for insurance purposes can best be explained by reviewing the insurance policy itself. The "standard 165 lines"[3] of most policies (Figure 10.1) offers the following data. "The insured shall...furnish a complete inventory of the destroyed, damaged and undamaged property, showing in detail quantities, costs, actual cash value, and amount of loss claimed;..."

At this point, the machinery and equipment appraiser must understand that the insurance policy is a contract between the insured and the insurance company. Because of this, the appraiser can best serve the needs of the client by having full knowledge of the standard 165 lines and any other riders or modifications attached to and part of the policy. An example of a rider to a policy is for a replacement cost policy. This would modify the 165 lines and the references to *actual cash value*. It is highly important that the client and insurance adviser have the insurance company accept the insurable replacement cost appraisal as the basis of the insurance policy and strike any reference to actual cash value. By so doing, the insured will be in as secure a position as possible. This will preclude any problems which might arise if the insurance company were to invoke use of the phrase under the heading, Company's Options, which states "...to repair, rebuild or replace the property destroyed or damaged with other of like kind and quality...".

A point of some concern arises with the insurance term *actual cash value*. According to all data available at this time, actual cash value is defined as replacement cost less physical depreciation by some sources and replacement cost less physical depreciation and obsolescence by

---

[3]It should be noted that, depending on the state, the number of lines can vary.

Insurable Value

Concealment, fraud This entire policy shall be void if, whether before or after a loss, the insured has wilfully concealed or misrepresented any material fact or circumstance concerning this insurance or the subject thereof, or the interest of the insured therein, or in case of any fraud or false swearing by the insured relating thereto.

Uninsurable and excepted property. This policy shall not cover accounts, bills, currency, deeds, evidences of debt, money or securities; nor, unless specifically named hereon in writing, bullion or manuscripts.

Perils not included. This Company shall not be liable for loss by fire or other perils insured against in this policy caused, directly or indirectly, by: (a) enemy attack by armed forces, including action taken by military, naval or air forces in resisting an actual or an immediately impending enemy attack; (b) invasion; (c) insurrection; (d) rebellion; (e) revolution; (f) civil war; (g) usurped power; (h) order of any civil authority except acts of destruction at the time of and for the purpose of preventing the spread of fire, provided that such fire did not originate from any of the perils excluded by this policy; (i) neglect of the insured to use all reasonable means to save and preserve the property at and after a loss, or when the property is endangered by fire in neighboring premises; (j) nor shall this Company be liable for loss by theft.

Other insurance. Other insurance may be prohibited or the amount of insurance may be limited by endorsement attached hereto.

Conditions suspending or restricting insurance. Unless otherwise provided in writing added hereto this Company shall not be liable for loss occurring
   (a) while the hazard is increased by any means within the control or knowledge of the insured; or
   (b) while a described building, whether intended for occupancy by owner or tenant, is vacant or unoccupied beyond a period of sixty consecutive days; or
   (c) as a result of explosion or riot, unless fire ensue, and in that event for loss by fire only.

Other perils or subjects. Any other peril to be insured against or subject of insurance to be covered in this policy shall be by endorsement in writing hereon or added hereto.

Added provisions. The extent of the application of insurance under this policy and of the contribution to be made by this Company in case of loss, and any other provision or agreement not inconsistent with the provisions of this policy, may be provided for in writing added hereto, but no provision may be waived except such as by the terms of this policy is subject to change.

Waiver provisions. No permission affecting this insurance shall exist, or waiver of any provision be valid, unless granted herein or expressed in writing added hereto. No provision, stipulation or forfeiture shall be held to be waived by any requirement or proceeding on the part of this Company relating to appraisal or to any examination provided for herein.

Cancellation of policy. This policy shall be cancelled at any time at the request of the insured, in which case this Company shall, upon demand and surrender of this policy, refund the excess of paid premium above the customary short rates for the expired time. This policy may be cancelled at any

Figure 10.1. The "standard 165 lines," which are included in most insurance policies and set forth the basic conditions governing the insurer and the insured.

time by this Company by giving to the insured a five days' written notice of cancellation with or without tender of the excess of paid premium above the pro rata premium for the expired time, which excess, if not tendered, shall be refunded on demand. Notice of cancellation shall state that said excess premium (if not tendered) will be refunded on demand.

Mortgagee interests and obligations. If loss hereunder is made payable, in whole or in part, to a designated mortgagee not named herein as the insured, such interest in this policy may be cancelled by giving to such mortgagee a ten days' written notice of cancellation.

If the insured fails to render proof of loss such mortgagee, upon notice, shall render proof of loss in the form herein specified within sixty (60) days thereafter and shall be subject to the provisions hereof relating to appraisal and time of payment and of bringing suit. If this Company shall claim that no liability existed as to the mortgagor or owner, it shall, to the extent of payment of loss to the mortgagee, be subrogated to all the mortgagee's rights of recovery, but without impairing mortgagee's right to sue; or it may pay off the mortgage debt and require an assignment thereof and of the mortgage. Other provisions relating to the interests and obligations of such mortgagee may be added hereto by agreement in writing.

Pro rata liability. This Company shall not be liable for a greater proportion of any loss than the amount hereby insured shall bear to the whole insurance covering the property against the peril involved, whether collectible or not.

Requirements in case loss occurs. The insured shall give immediate written notice to this Company of any loss protect the property from further damage, forthwith separate the damaged and undamaged personal property, put it in the best possible order, furnish a complete inventory of the destroyed, damaged and undamaged property, showing in detail quantities, costs, actual cash value and amount of loss claimed; and within sixty days after the loss, unless such time is extended in writing by this Company, the insured shall render to this Company a proof of loss, signed and sworn to by the insured, stating the knowledge and belief of the insured as to the following: the time and origin of the loss, the interest of the insured and of all others in the property, the actual cash value of each item thereof and the amount of loss thereto, all encumbrances thereon, all other contracts of insurance, whether valid or not, covering any of said property, any changes in the title, use, occupation, location, possession or exposures of said property since the issuing of this policy, by whom and for what purpose any building herein described and the several parts thereof were occupied at the time of loss and whether or not it then stood on leased ground, and shall furnish a copy of all the descriptions and schedules in all policies and, if required, verified plans and specifications of any building, fixtures or machinery destroyed or damaged. The insured, as often as may be reasonably required shall exhibit to any person designated by this Company all that remains of any property herein described, and submit to examinations under oath by any person named by this Company, and subscribe the same; and, as often as may be reasonably required, shall produce for examination all books of account, bills, invoices and other vouchers, or thereof if originals be lost, at such

Figure 10.1. *(Continued)*

Insurable Value **157**

> reasonable time and place certified copies as may be designated by this Company or its representative, and shall permit extracts and copies thereof to be made.
>
> *Appraisal.* In case the insured and this Company shall fail to agree as to the actual cash value or the amount of loss, then, on the written demand of either, each shall select a competent and disinterested appraiser and notify the other of the appraiser selected within twenty days of such demand. The appraisers shall first select a competent and disinterested umpire; and failing for fifteen days to agree upon such umpire, then, on request of the insured or this Company, such umpire shall be selected by a judge of a court of record in the state in which the property covered is located. The appraisers shall then appraise the loss, stating separately actual cash value and loss to each item; and, failing to agree, shall submit their differences, only, to the umpire. An award in writing, so itemized, of any two when filed with this Company shall determine the amount of actual cash value and loss. Each appraiser shall be paid by the party selecting him and the expenses of appraisal and umpire shall be paid by the parties equally.
>
> *Company's options.* It shall be optional with this Company to take all, or any part, of the property at the agreed or appraised value, and also to repair, rebuild or replace the property destroyed or damaged with other of like kind and quality within a reasonable time, on giving notice of its intention so to do within thirty days after the receipt of the proof of loss herein required.
>
> *Abandonment.* There can be no abandonment to this Company of any property.
>
> *When loss payable.* The amount of loss for which this Company may be liable shall be payable sixty days after proof of loss, as herein provided, as received by this Company and ascertainment of the loss is made either by agreement between the insured and this Company expressed in writing or by the filing with this Company of an award as herein provided.
>
> *Suit.* No suit or action on this policy for the recovery of any claim shall be sustainable in any court of law or equity unless all the requirements of this policy shall have been complied with, and unless commenced within two years next after inception of the loss.
>
> *Subrogation.* This Company may require from the insured an assignment of all right of recovery against any party for loss to the extent that payment therefor is made by this Company.

Figure 10.1. *(Continued)*

other sources. Regardless of the ability of the insurance industry or the courts to be in agreement on this point, it is our opinion that the definition of insurable value depreciated included within this text best fits the needs of all concerned.

Methodology

The insurance appraisal itself demands classification of property, description and location of each item, and verifiable costing or valuation in order to meet the requirements of the insurance policy and to be of most benefit to the client. The proper procedure to follow for the classification, description, and estimate of replacement cost is covered elsewhere in this text.

Requirements for Detailed Appraisal Report

The reasons for the amount of detail in this type of appraisal report are multiple. Most important is the case of a partial loss, such as the destruction of only a portion of an industrial building or complex. A detailed appraisal showing proper classification of each item with an accompanying description, location (building, floor, department, etc.), and valuation is a most important tool in the preparation of proof of loss papers. A second reason for these requirements is that different classes of property may be insured at different rates. The detailed appraisal report, therefore, provides the insured with the greatest ability to be able to save money in the form of reduced premiums. Further, the appraisal report will attest to the fact that the insured has done the utmost to meet the obligations of the insurance contract.

In some instances, any of the three approaches to value can be used in the valuation of machinery and equipment. In the case of insurance appraisals, the cost approach and the market approach are the methods that are utilized.

Procedure for Determining Insurable Replacement Cost

Replacement cost new is defined and explained earlier in this text. In order to arrive at the insurable replacement cost, the appraiser has to make deductions from the replacement cost new for any exclusion which may be listed in the policy and which exist at the subject. Examples of excludable items are machine foundations; the cost of associated excavations, backfill, etc.; and below ground piping. A statement should be made to the effect that the exclusions are indicated as a convenience for the client and should not be construed as meaning that they are not liable to be damaged nor that they should be excluded from coverage. The determination as to which items are to be excluded should be made by the client and the insurance adviser.

Procedure for Determining Insurable Value Depreciation

It is under the heading of insurable value depreciation that the items of property under consideration may require individualized treatment. In some instances, such as for items which may still be available from a manufacturer in the same or similar manner, depreciation for insurance purposes can be measured by the following procedure. The appraiser, after a personal inspection of the item, its manner of usage, maintenance records, and the like, must analyze the condition of the item. Based on this analysis, an effective age can be estimated. It is a simple mathematical procedure, from that point, to provide a depreciation percentage by dividing the effective age by the estimated serviceable life.

For example, let us consider an item that was built and first put into service in 1970. The item was originally estimated to have a serviceable life of 12 years. Discussions with the manufacturer indicate that the item is still available in much the same manner. Personal inspection with investigation of usage and maintenance procedures lead the appraiser to determine that the apparent effective age of the item, regardless of its chronological age, is five years. Five years (effective age) divided by 12 years (serviceable life) indicates a depreciation factor of 41.7 percent. This factor, when applied to the insurable replacement cost leaves an amount which is termed *insurable value depreciated:*

Replacement cost new	$19,000.00
Less: 41.7 percent depreciation	−7,923.00
Insurable value depreciated	$11,077.00

In the case of items which are no longer manufactured in the same way or which have become obsolete, the appraiser has to consider the availability of the same or similar items in the used machinery marketplace. This would be in keeping with the standard 165 lines where, at line 145, the phrase "like kind and quality" is used. Items which are to be included in this manner should be so indicated by the use of an asterisk and an accompanying explanation. Naturally, it is the installed cost of the used item that will be included. Further detail with regard to this approach can be found in "The Cost Approach" in Chapter 8.

Coinsurance

Another point which must be addressed is coinsurance. Coinsurance is a system whereby a premium reduction is granted to an insured who agrees to carry an amount of insurance equal to or greater than a stated percentage of the insurance value depreciated of the subject property. An insured person who does not maintain the proper level of coverage becomes a *coinsurer* of a portion of the loss. It is in the event of a partial loss that coinsurance may come into play, since the face value of the policy will be paid in the event of a complete loss.

A simple manner in which coinsurance may be understood is by use of the following formula:

$$\frac{\text{Amount of coverage}}{\text{Amount of coverage required}} \times \text{loss} = \text{settlement}$$

At no time, however, will the settlement exceed the face value of the policy.

Although coinsurance is a matter for discussion between the insured and the insurance adviser, the appraiser should have an understanding of the subject. A detailed explanation can be found in the chapter titled "Co-Insurance—What It Is and How It Works," *Insurance Valuations*, Monograph #4, July 1971, American Society of Appraisers.

Conclusion

Appraisals for insurance purposes are more than an inventory. To provide the insured with the best possible base from which to place insurance, the appraiser should become as expert as possible with the individual insurance policy and the items under consideration. By so doing, the appraiser may well be of great assistance in the formation of a single language, set of definitions, and format for the insurance industry.

11
Scrap/Salvage

David M. Graham, ASA
Greenbank, Washington

Scrap and/or salvage value of machinery and equipment is often equated with junk value and therefore has no realistic basis for inclusion in most appraisals. On occasion, however, there may be a valid reason to determine the scrap and/or salvage value. Primarily, the reason is to estimate the recovery from the disposal of machinery and equipment which is surplus, obsolete, incomplete, or damaged. Any of these conditions would render an item useless to an operating plant on physical, functional, or economic grounds. This does not necessarily mean the item is worth nothing, albeit the value would likely be lower than for any other purpose of appraisal.

Definitions of Scrap and Salvage Values

While the terms *scrap* and *salvage* are often used interchangeably with an implied equivalent meaning, there is a definite difference between the two terms.

Scrap value primarily means the tonnage or pound price of basic, recoverable metals, which may include iron, steel, stainless, copper, aluminum, or titanium. (While not common, the term might also apply to lumber or other basic materials.)

Salvage value means the value of recoverable machines or equipment or parts thereof including, but not limited to, base castings, gears, shafts, or other mechanical components such as controls, gauges, valves, pipe fittings, or electrical parts.

Determining Scrap Value

The problem of determining scrap value is not very complex. First, the appraiser estimates the pounds (or tons) of each metal to be scrapped. Second, the appraiser ascertains from local scrap metal buyers the current price being paid for the metals in question. With respect to these steps, however, there are important considerations to be made in order to determine properly the net value to the client. Net, of course, is the only value of concern to the client.

> *Net* is the realization after deducting any costs which may be incurred to prepare the scrap for sale.

Therefore, the appraiser must determine if there will be any costs involved in separating the different metals from each other. A metals buyer will either require metals to be separated or will offer a lower price if it has to be done by the buyer. Also, the appraiser must find out if the buyer's price includes pickup and hauling to the yard or if the client must load and haul to the buyer's yard, thus creating a cost to weigh against the buyer's price.

Determining Salvage Value

The problem of estimating salvage value is somewhat more complex. First, the appraiser must decide if the carcass of a unit is sufficiently sound and capable of being rebuilt to a useful condition on an economically practical basis. Alternatively, an appraiser may decide the carcass is capable of being utilized by modification in the custom manufacture of a machine for some use other than original design. The decisions referred to here may be best made by consultation to ascertain if the client has any plans for future use. Furthermore, it may be relevant to consider parts value, i.e., cannibalizing machines for parts that may be useful to the client as replacement spare parts for in-house future use or for sale to a dealer who may wish to retain a parts inventory for customers.

To estimate the value of salvage, consider the following:

- If a machine carcass were to be rebuilt to a useful condition or used as the starting point for the custom manufacture of a different use machine, what would be the cost required to purchase an equivalent carcass from a used machine or junk dealer? If no reuse is anticipated, the carcass might be considered for a parts value.

- If parts value for in-house use is relevant, what would be the cost to acquire equivalent parts from a parts dealer, liquidator, or junkyard?
- If parts value for sale to a dealer or liquidator is relevant, what would the dealer or liquidator pay for such parts?

As with scrap, salvage values should be net. Therefore, consider the costs to cure as an offset to value as recovered. This means the cost to strip, clean, and condition the carcass or parts for future in-house use. Alternatively, it means the cost to separate components and deliver, if required, to the purchaser.

In the category of salvage, it is entirely possible that even a superficial inspection by the appraiser and brief consultation with the client as to future plans might, from an economically practical point of view and without lengthy consideration, reduce salvage to scrap.

Scrap/Salvage Analysis

Occasionally, an appraiser might have a client who for some reason may have the need for a scrap/salvage analysis of an entire plant. If plant demolition were the factual objective, then the owner would probably employ a demolition company to remove the plant and negotiate with them whatever recovery was possible. On the other hand, the appraiser may be involved if the analysis is needed, but the plant is not to be actually destroyed. This would most likely be a hypothetical appraisal.

To deal with this salvage question, however, we must suppose that all machinery and equipment, furniture, and fixtures which are marketable are considered disposed of under a fair market value for removal and/or liquidation appraisal. The results reduced by the cost of such dispositions could be considered a net salvage value. There would then remain for the true salvage analysis such things as structures, foundations, piping, wiring, earthwork, and miscellaneous and surplus items.

These remaining quantities should then be considered as described above, first as to salvage and then as to scrap value if necessary. Again, net is all that is important, and it must be recognized that net could well be negative with respect to any item, system, or even the whole plant. Remembering that net is the recovery after deducting the costs of recovery and that our analysis requires considering reducing the plant to a bare ground condition (including foundation removal), we note the following likelihoods:

- Structures may or may not have a net salvage value.
- Foundations will have a negative value.
- Piping may or may not have a net value, depending on the content and marketability of valves, instruments, etc.
- Wiring may or may not have a net value, depending on the content and marketability of panels, switches, transformers, etc.
- Earthwork will have a negative value.
- Miscellaneous and surplus may or may not have a net value depending on content and marketability. N.B. Some plants may have very sizable net recovery from catalysts, plating solutions, and recoverable oil.

The salvage analysis on an entire plant usually involves the consideration of the demolition of buildings. While buildings may not seem a proper consideration in a machinery and equipment analysis, the subject is included since the entire plant example would be incomplete without them. Furthermore, it is recognized that in many plants special purpose buildings are so interrelated to equipment that total separation of analyses is quite impractical. The appraiser may have to obtain a proposal from a demolition contractor for this part of the study.

Alternatively, some guidance is available in such publications as *Building Construction Cost Data* (Robert Snow Means Company, Inc., 100 Construction Plaza, Kingston, Massachusetts 02364) or *National Construction Estimator* (Craftsman Book Company, 6058 Corte del Cedro, Box 6500, Carlsbad, California 92008-0992). Whether building demolition is estimated based upon a contractor's proposal or an approximation of the client's in-house costs, it is probable that no net recovery will exist. It is more likely that costs will significantly outweigh any possible recovery.

It is therefore apparent that a salvage (including scrap) analysis of an entire plant is really a combination of values from several analyses.

Sources of Scrap/Salvage Value Information

Sources of information useful in the estimating of scrap and/or salvage values would include:

Scrap metal buyers

Exotic metal buyers

Used machine dealers

Junk dealers
Liquidators
Used parts dealers
Truck rental agencies
Client's in-house costs for yard labor
Demolition service companies

It is possible that some of the above information, once obtained, could be computerized for future use. It is doubtful, however, that a database would be of substantial use, because the need for serious scrap/salvage analysis is rather rare.

12
Value-in-Use versus Value-in-Exchange

Leslie H. Miles, Jr., ASA
CEO, MB Valuation Services, Inc., Dallas, Texas

It is necessary to know the type of value desired before the appraiser can begin his or her report. There are three primary values that are referred to for machinery and equipment: liquidation, fair market, and replacement. The American Society of Appraisers has taken these primary references and formed terms which are separately defined and which incorporate the primary heading. Chapters 4, 8, and 9 within this text explain these concepts. It is recommended that the reader review the definitions for liquidation value, orderly liquidation value, liquidation-in-place, fair market value, fair market value-in-place and value-in-use, fair market value-in-place and value-not-in-use, reproduction cost, and replacement cost. The three chapters which incorporate these definitions explain the value application and give a clarification by expanding on that value which has been defined. Understanding the concepts of value makes it simple to follow this chapter.

Definition of Value-in-Use

Value-in-use is the amount, expressed in dollars, that the property is worth to the user and may be based on the items' capability of producing a product or part of a product as a measured percentage of profit; a contribution to other items, which is difficult to measure; items of a promotional nature; or value to the user on a required personal basis.

Since the income approach applied to a particular piece of equipment is thoroughly explained in Chapter 8, there is no need to reiterate the method of its application for value-in-use. Equipment not easily measured in production would be such types of support equipment as storage racks, general plant furniture, and office furniture. Promotional items could be such things as boats, airplanes, demonstration equipment, prototypes, and research and development equipment. Items which may be valuable only to the user could be those which would be considered luxuries such as a Rolls Royce automobile, jewelry, and personal decor items.

To the user, who has to purchase all the above items, the value is at the upper end of a range that might be set for any like item. The easiest example is a relationship to a piece of jewelry, such as an 18-carat-gold Rolex "President's" watch which may cost in excess of $8000. The jewelry appraiser and the owner would consider the original cost as the value of the watch, at a minimum, and, in all probability, the new cost at any given point in time. Of more interest to the machinery appraiser, a user may apply that same logic to a computer, numerically controlled turning center, which is only one generation behind the latest model since it has all the capabilities of a new one, including all additional upgrades, which makes it the same in the user's mind.

Value-in-Exchange

Value-in-exchange may disappoint the seller of the 18-carat-gold watch, assuming it is the same individual described in the above paragraph. The newspapers indicate offerings of this watch, due to supply and demand, between $5000 and $6,000. In that example it could be a very simplistic understanding of value-in-use versus value-in-exchange. The buyer of a CNC turning center may or may not have the warranty to consider in the purchase but will make other considerations. The buyer of the turning center would figure reinstallation cost and possibly, depending on the situation, even the cost of removal. With computer technology changing at an ever increasing pace, CNC controls may or may not be upgradable but, historically, take a high depreciation as new equipment replaces the old. For whatever reasons, the new buyer will pay an amount that is fair in the marketplace, commensurate with a particular demand or need, and only after all cost burdens are considered.

In some cases, although very few, it is possible for value-in-exchange to be greater than value-in-use to a particular user. This is not normally something that can be anticipated by an appraiser, but it occasionally happens where demand exceeds supply or availability. In the middle to late 1960s, there were auction sales of Bridgeport mills at or above new cost for the very reason of supply and demand. These mills would bring a price approaching $5000 to $6000 at a time when they only cost $3500 new. This was prior to the influx of foreign competition with such names as Induma, Index, and Sajo, some of which almost duplicate the specifications of the very popular and standard piece of equipment known as the Bridgeport mill.

In the 1970s there were positions taken (orders) on Learjets, Falcons, and other aircraft which were later sold for a higher price on delivery owing to the rate of inflation from the time of purchase to the date of delivery. It should be noted in this example that the purchase price was at the current new cost of a new aircraft even though above the purchase price of a position taken at an earlier date. In this case, the sales motivator was that of immediate delivery rather than a six- to eight-month wait on a new order. This example is used to show that there is no confusion regarding inflation's effect on future sale prices at greater than the original price. The example is also the exception rather than the rule, which should be understood in making relationships or measurements to historical sales (comparables). In reality, the aircraft had an equal value-in-use and value-in-exchange because it was new and the price was the current new price at the time of delivery. This is an entirely different situation from one in which an inflated price is caused by supply and demand, such as the case of the Bridgport mill. In the late 1970s, there were prices paid for used drilling rigs of one and a half to two times replacement cost new for an equivalent unit. The demand was there for "holes to be punched in the ground" without the supply of equipment to do it. This effect trickled down throughout the drilling industry and affected support equipment such as drill pipes, drill collars, and pump jacks.

Value-in-Use versus Value-in-Exchange

To the appraiser, there is little or no confusion between a value-in-use and a value-in-exchange. In many cases there is confusion between the owner of an item and the potential buyer, which the appraiser clears up in the valuation study. Before an appraisal project begins, the paying client should be clear on the concept of value and how it is defined. Value-in-use can be the same as value-in-exchange when the purchaser is to become a like user and, therefore, pays that price for the equipment or entity. It might be easier to say that a value-in-use is not as probable in recovery as a value-in-exchange. A purchaser may buy an entire plant of equipment but may be doing so for some form of liquidation. In this case, the price being paid would be based on a value-in-exchange and, in all probability, somewhere below fair market value on a removal basis in order to make an anticipated profit.

This reference to fair market value is based on dollar value gleaned from the resale market to a user. There are times when fair market value may be applied to that which a used machinery dealer pays in the resale market.

There are three concept applications referred to in the first paragraph, and it is assumed that the reader generally understands these terms as they are covered in other chapters in the textbook. These concept applications can be confusing when they are relating to value-in-use versus value-in-exchange. A review is offered here to clarify this point which was emphasized at the beginning of the chapter and is reiterated with emphasis at this time.

Discussion of Liquidation Concepts Related to Value-in-Exchange and Value-in-Use

Liquidation concepts are the most easily understood for value-in-exchange. However, the liquidation concept can use an in-place application that may be misunderstood. The concept of *liquidation value-in-place* approaches *value-in-use* under a qualified set of circumstances. It assumes that all machinery and equipment, and in some cases real estate and its related improvements, will be sold intact for the purpose intended, even after failure. A purchaser would prudently base a price to be offered on a recapture of investment, even after failure and, therefore, would apply a risk factor in the development of the offering price. It is sometimes assumed that liquidation value-in-place is a discounted fair market value and, although typically lower than fair market value, is not necessarily computed by that method.

Liquidation value-in-place is often confused with fair market value because of the client's perception as to what must be considered for in-place/in-use. A client comparison may still consider a cost factor plus installation in computing any discount as there is very little weight given to the product line or customer list. Regardless of the understanding, liquidation value-in-place considers a value-in-use with an application made after a hypothetical failure. At a later time, a user/buyer may not be available, even as a speculator, at the price incentive indicated in the appraisal study. The reasons for this may be a change in the marketplace, a failure due to product line or acceptance of same, cost of money, change of equipment condition, general industry economics, or various other reasons. At the point in time in which there is no buyer owing to the reasons stated, value-in-use is not the same as value-in-exchange.

Fair Market Value Relationship to Value-in-Exchange and Value-in-Use

Fair market value is analogous to liquidation value but without any failure concepts being applicable. *Fair market value-in-place-in-use* assumes the equipment would be used for the purpose intended where it is, no differently than might be considered by the current owner. However, fair market value removed or fair market value-in-place-not-in-use can better be considered value-in-exchange. Fair market value relies on the marketplace in determining a willing buyer's purchase price that is also fair to the seller when considering that the equipment would be removed by the purchaser to be placed elsewhere. The fair market value-in-place-not-in-use considers a failed facility in which the equipment remains idle and/or is to be purchased for use in an alternative or adaptable manufacturing process.

An example would be a machine shop that manufactured drill bits and is purchased by a company that wishes to manufacture pistons. The new purchaser may consider that most, if not all, of the equipment can be utilized in the new product line but that certain adaptations would be removed and/or installed for the new product line. Fair market value and fair market value-in-place-not-in-use are normally considered value-in-exchange, whereas fair market value-in-place-in-use better approaches value-in-use. It might be well to say that if the purchaser of a company captures the product line and wishes to continue the present operation, the fair market value approach is consistent with value-in-use, whereas removal or alteration of product line better approaches value-in-exchange.

Replacement Concepts Related to Value-in-Exchange and Value-in-Use

Replacement concepts relate more to insurance-type studies and may be either value-in-use or value-in-exchange, depending on the situation. Most users do not consider that their equipment has value equal to its new replacement cost but, in many instances, consider the depreciated replacement cost as equal to fair market value. Physical depreciation is the only obsolescence factor applied to a piece of equipment when making insurance considerations and, therefore, may be value-in-use rather than value-in-exchange.

If other types of obsolescence factors, such as economic and functional, are applied, the value may then approach value-in-exchange subject only to the condition of the market. If a disaster were to occur, the owner of the equipment would be replacing the machinery and therefore would consider value-in-use a proper dollar recovery. It is possible that the used equipment market may allow repurchasing of like equipment at a different price than was arrived at by some percentage application to replacement cost new. This marketplace purchase at replacement would then be value-in-exchange.

When Value-in-Use Is Not Comparable to Value-in-Exchange

Let us consider the hypothetical manufacturer of pump parts in which labor-intensive equipment is diminishing the company's competitive edge on a yearly basis. Unless something is done immediately, the company will be in a losing position with no hope of recovery. A consultant is hired who designs and can produce a specialty machine that can save the company from disaster if ordered immediately. There is only one machine that can be found for conversion and then only at a price not consistent with the average market. However, after all costs are considered, it is a bargain at twice the price, owing to the particular need of the company. The price is paid, with additional money for adapted features, which cost the same as the machine's base price.

In this example, we assume that the machine works, saves the company from failure, and will continue to make money as long as the product is made and demand is at or above known sales at the time of its acquisition. The price that has been paid by this company is value-in-use and will continue to be realistically a fair value to the user throughout its economic life. It is also assumed that no other company manufactures the product in the same manner as the current user; there would only be a purchaser if the company's business and product line could be captured. If the manufactured product were eliminated or if an alternative method of manufacturing were found, a new purchaser for the business would not consider, in all probability, the original purchase price a fair value for this machine. In any event, a piecemeal sale of this machine in the open marketplace would not attract a purchaser, except for the base machine and only then after restoration to its original basic design. In this example, there is a high disparity between value-in-use and value-in-exchange.

Purchasing of equipment is typically done by one of two categories of buyers, *user buyers* or *investor buyers*. In the above example the user buyer is paying a value-in-use rather than a value-in-exchange. A user buyer may pay value-in-exchange when time is not of the essence and a fair and equitable price can be paid through the open marketplace consistent with what that marketplace would bear. An investor, in most instances, would be an individual or company, such as a machinery dealer or leasing company. A prudent investor typically pays at or below the user market prices.

When Value-in-Use Can Equal Value-in-Exchange

The profitability of equipment is dependent on value-in-exchange, and therefore, an appraiser must anticipate that a purchaser or user is available. The remarketing of equipment has certain guidelines within which the prudent appraiser must stay. These guidelines must allow a reasonable value indicator when relating to a standard marketplace in order to sell the piece under question. There must be considered a probability of sale when the equipment is exposed to the marketplace. Value-in-use anticipates a sale with the restriction of like need by a new purchaser. That need must be specific and exact to the requirements of the current user. Because of this, value-in-use and value-in-exchange can be identical. However, concepts of value without the term *in-use* must take into consideration the highest probability of sale as well as what occurs in the marketplace.

Summary

If there is an understanding of the difference between value-in-use and value-in-exchange, it is necessary to know only how to determine that value. Chapters 8 and 9 cover value determination under many value situations with some references to *in-use*. However, an appraiser must not become confused in making value judgments. The easiest way to keep things in perspective is to consider value-in-use as one in which all equipment, adapted or otherwise, has value to its present manufacturing process, which should continue. This, then, is value to the user and more than likely would also be applied to a new purchaser. Value-in-exchange, on the other hand, may consider special adapted equipment as having limited marketability and, therefore, higher depreciation in the average marketplace. Value-in-exchange must consider what typically happens in the marketplace and the probability of resale considering adaptability, standard equipment demand, all forms of obsolescence, and other causes and effects. If value-in-use and value-in-exchange are equal, there should be justification for it.

This chapter is not intended to teach appraisal application under various concepts of value but rather to explain remarketing of equipment. An appraiser must anticipate what will happen in the marketplace and must understand value-in-use versus value-in-exchange.

13
Appraisal Report Content

George D. Sinclair, FASA
*President, Keystone Appraisal Company
Philadelphia, Pennsylvania*

The report that an appraiser submits to a client is the finished evidence and the result of the investigation, research, and analysis. It is the communication of the appraiser's findings and the answer to the question that the appraiser was retained to answer.

It is the only benchmark by which the appraisal-buying public, the client, can rate the quality of an appraiser's work and judge someone as an appraiser. The reader of the report will never know of the time and effort that went into the appraisal unless the report sets forth in a logical and readable manner the factors taken into consideration in arriving at the final conclusion of value.

The machinery and equipment appraiser has a duty not only to the appraisal-buying public but also to the appraisal profession to make the report an effective instrument that conveys his or her findings in a professional manner.

The Principles of Appraisal Practice and Code of Ethics of the American Society of Appraisers outlines in detail the appraiser's responsibility and what is required in the preparation of an appraisal report. No one should attempt to make an appraisal, or submit an appraisal report, without a complete understanding of these requirements.

The reader is referred to Chapter 14 of this text and to *The Princi-*

ples of Appraisal Practice and the *Code of Ethics* of the American Society of Appraisers.

The machinery and equipment report must meet a uniform standard that is professional and in compliance with good appraisal practice and sound appraisal theory. The length and content of a machinery and equipment report varies, and the amount of data and details incorporated depends on the estimate of value or estimates of values required and on the complexity of the assignment.

Minimum Requirements of an Appraisal Report

However, there are certain minimum requirements that are mandated for all machinery and equipment reports. In fact, the North American Conference of Appraisal Organizations has endorsed a summary of suggested minimum appraisal standards that is applicable to all appraisal reports.[1]

These minimum appraisal standards state that there are four major steps involved in an appraisal (which is a research project):

1. Preparation of an outline, a plan, or blueprint for action
2. Assembling and classifying the material for analysis, the data program
3. Applying the tools of analysis, the analytical techniques, and the approaches for interpreting that data
4. Applying the skilled mental labor necessary to unite these ingredients in a meaningful conclusion

Letter of Transmittal

The first portion of an appraisal to be seen by any reader of a machinery and equipment appraisal report is, of course, the letter of transmittal.

This letter should set forth a concise summary of the factual data, analyses, and conclusions that are contained in the body of the report. It must also state the type of property being valued, an identification of the type of value, and the date of valuation.

---

[1]*Valuation*, vol. XXX, no. 1, December 1984, American Society of Appraisers.

The signature of the appraiser responsible for the value conclusion set forth in the report is required. There are occasions when more than one appraiser will work on a given assignment. When this is the case, each appraiser should sign the letter of transmittal or have it noted in other appropriate sections of the report that each appraiser accepts the responsibility for the value conclusion as stated in the report.

There are no standard letters of transmittal. Each appraiser may have his or her own form or standard letter. Also, owing to the various estimates of value that are required of a machinery and equipment appraiser, no standard form letter will fill all the changing market demands for the various value estimates that may be required.

However, a sample letter of transmittal that might fill the needs of various appraisal requirements is shown in Figure 13.1.

The Report

An appraisal is an opinion of an estimate of value, and it must always be backed up by a complete written report or memorandum. It should

---

Name of Person, Company, etc., to Whom Report Is Directed

RE: Identification of Facility Being Appraised
     Address

Dear Mr. Client:
In accordance with your request, we have prepared an appraisal for the purpose of arriving at an opinion as to the fair market value-in-place-in-use of the machinery, dies, tools, plant furniture and equipment, and office furniture and equipment, located in your facility at
Based on the data, analysis, and conclusions set forth in the detailed report which follows, in our opinion the fair market value-in-place-in-use of the appraised items, as of _____, is as follows:
$_____
In this report, we have made no investigation of the property ownership, nor have we taken into consideration any encumbrances which may be against it. Our work has to do only with the estimate of values.

Respectfully submitted,

MACHINERY AND EQUIPMENT APPRAISAL COMPANY

Your Appraiser, President

Figure 13.1. Sample letter of transmittal.

never be an offhanded expression of value. The exact order of the items included in a report may change depending on many factors, but all should be included, either with a separate heading or as part of one of the other sections.

The type of information or data which forms each section should include, but not necessarily be limited to, those listed in the following sections.

Identification of Property

The name of the company or individual represented as owning the asset(s) and the location or locations of the appraised property should be listed.

It is important to specify the asset(s) being valued. Are you appraising only certain specific items of machinery and equipment, or are you appraising all the machinery and equipment? Does your report include office equipment, licensed vehicles, dies, or any special assets?

Examples of how the identification of property may be worded should include the locations(s) where all the machinery and equipment is being valued. An example of how this information may be presented is stated below:

> **Identification of Property**
>
> The property appraised in this report consists of the machinery, equipment, and fixtures of XYZ Corporation, located at 210 Main Street, Anytown, PA. It also includes those items of machinery presently located in Mytown, NY, which are to be moved to the Anytown, PA, facility.
>
> The appraised items are metalworking machines such as vertical mills, grinders, planers, dies, tooling machines, machining centers, and inspection equipment which includes surface gauges, protractors, and comparators. Office furniture, office machines, and rolling stock are also included in the appraisal.

Where only a specific line of machinery and equipment is being valued, this section of the report may read:

> **Identification of Property**
>
> The property appraised in this report consists only of certain items of machinery and equipment presently located in the Press Department of XYZ Corporation, 100 South Street, Bigtown, NY, as specified in the letter of authorization to make the appraisal.

Purpose of the Appraisal

The value or values being estimated must be defined. An appraisal must be based on an understanding of the value as defined and the valuation process set forth in the report. The date or dates of the value estimate must be stated.

It is not unusual for a machinery and equipment appraisal report to express an opinion as to more than one value. In fact, many machinery and equipment appraisals require the appraiser to express an opinion as to three or four different value estimates. These may include value-in-place-in-use, fair market value, and estimated liquidation value.

Examples of how this could be expressed would include, but would not be limited to, the material in the following sections.

Valuation for Market Value-in-Place-in-Use

Purpose of the Appraisal

The purpose of this appraisal is to estimate the market value-in-place-in-use of the fee simple interest, unencumbered, of the subject machinery, equipment, and inventory under conditions prevailing as of month, date, year.

Market value-in-place-in-use for the purpose of this report is defined as "The fair market value of an item, including installation and the contribution of the item to the operating facility."

This value presupposes the continued utilization of the item in conjunction with all other installed items of machinery and equipment which are included in the appraisal.

A statement such as the following could be added, if required, to clarify the appraiser's concept of this value.

Regardless of the exact wording of the definition, *market value-in-place-in-use* contemplates the consummation of a sale and the passing of full title from seller to buyer under conditions whereby:

Buyer and seller are free of undue stimulus and are motivated by no more than the reactions of typical owners.

Both parties are well advised or well informed and act prudently for what they consider their own best interest.

A reasonable time is allowed to test the market.

Payment is made in cash or in accordance with financing terms available in the community for the property type in its locale.

Market Value for Off-Site Use. There are times when a market value for offsite use is required, and this must be distinguished from market value-in-place-in-use.

Purpose of the Appraisal

The purpose of this appraisal is to estimate the fair market value for offsite use of the fee simple interest, unencumbered, of the subject property under conditions prevailing as of month, date, year.

Fair market value for offsite use is defined as that value to a user for the item of machinery or equipment, after comparison of subject items with similar items sold in the market, used machinery dealers, and/or individual transactions. Usually the buyer is responsible for cost of moving and transporting all items from site, and this should be stated.

Estimated Liquidation Value

There are three definitions of liquidation value that are in common use and that have been accepted as being applicable to the appraisal of machinery and equipment under liquidation conditions. Therefore, the purpose of an appraisal under liquidation conditions requires a clear definition as to the exact value being estimated. A sample wording of these three liquidation value definitions would then be:

Liquidation value is the estimated gross dollar amount which could be typically realized at a properly conducted public auction held under forced conditions and under present day economic trends.

Orderly liquidation value is the amount of gross proceeds which could be expected from the sale of the appraised assets held under orderly conditions given a period of time in which to find a purchaser(s) considering a complete sale of all assets, as is, where is, and all sales made free and clear of all liens and encumbrances.

Liquidation value-in-place is an amount in money which is projected to be obtainable considering the present marketplace assuming that an entire facility would be sold intact along with all related equipment necessary to make it viable. It further considers that fair market value, as normally defined, could not be obtained owing to restrictions of time and probable conditions of the business under forced sale conditions.

These definitions of liquidation value must also be clarified as to who is responsible for the cost of moving the item of machinery and repairing any damage to the real estate if that is a factor to be considered. If the assets must be sold within a specific time, 30 days to 90 days, six months to one year, that also must be stated.

The function of the appraisal relates to the character of the decision that is to be based on the value estimate expressed, and it should not be confused with the purpose of the appraisal.

A few examples of where an appraisal made to estimate market value-in-place-in-use (purpose) of machinery and equipment could have a different function follow:

1. Buy-sell agreement

2. Purchase price allocation
3. Condemnation
4. Collateral for a loan
5. Local taxes, where machinery and equipment are subject to local or state taxation
6. Charitable deduction

Regardless of its function, the purpose of the report is paramount. Market value-in-place-in-use is not different nor does it change according to the use to which the client will put the report. Market value-in-place-in-use is the same for tax purposes, or for a buy or sell agreement, as it is in condemnation where a government agency acquires the assets.

It must be noted at this point that the Tax Reform Act of 1984[2] has imposed stringent regulations regarding appraisals made to obtain charitable deductions. These regulations are applicable to any property donated, other than money or publicly traded securities, when the amount of deduction or reported appraised value exceeds $5000.

An analysis of the regulations required for an appraisal report to be considered as a qualified report is basically the same while in a different format as outlined in this section.

A summary of the information required to meet these IRS standards includes the following:

1. A description of the property in sufficient detail for a person who is not generally familiar with the type of property to ascertain that the appraised property is the property that was (or will be) contributed
2. In the case of tangible property, its physical condition
3. The date (or expected date) of contribution to the donee-charity
4. The terms of any agreement or understanding entered into (or expected to be entered into) by or on behalf of the donor that relates to the use, sale, or other disposition of the contributed property—including, for example, the terms of any agreement or understanding that
   a. Restricts (temporarily or permanently) the donee-charity's right to use or dispose of the donated property
   b. Reserves to, or confers upon, anyone (other than the donee-charity or an organization participating with the donee-charity

---

[2]Regulation 1.170A-13, J.D. 8002, *Federal Register,* vol. 49, no. 252, pp. 50660 and 50664, December 31, 1984.

in cooperative fundraising) any right to income from the donated property or possession of the property (including the right to vote donated securities, to acquire the property by purchase or otherwise, or to designate the person having the income, possession, or right to acquire), or

   c. Earmarks donated property for a particular use

5. The name, address, and taxpayer identification number (Social Security number) of the qualified appraiser (or appraisers) and—if the qualified appraiser (or appraisers) is a partner in a partnership, an employee of any person (whether an individual, corporation, or partnership), or an independent contractor engaged by a person other than the donor—the name, address, and taxpayer identification number (employer identification number) of the partnership or person who employs or engages the qualified appraiser (or appraisers)
6. The qualifications of the qualified appraiser (or appraisers) who signs the appraisal, including the appraiser's background, experience, education, and membership, if any, in professional appraisal associations
7. A statement that the appraisal was prepared for income tax purposes
8. The date (or dates) on which the property was valued
9. The appraised fair market value of the property on the date (or expected date) of contribution
10. The method of valuation used to determine the fair market value (e.g., the income approach, the market data approach, or the replacement cost-less depreciation approach)
11. The specific basis for the valuation, if any (e.g., any specific comparable sales transactions)
12. A description of the fee arrangement between the donor and appraiser

Description of the Machinery and Equipment

A narrative description of the machinery and equipment, or other assets being appraised, as to type, function, and its utility is required at this point.

It should be understood that this section does not replace the detailed inventory of appraised assets, but it is a narrative that gives the reader of the report a short summary of the assets and the appraisal problem to be solved.

If there are any known or observed encumbrances, limitations, or restrictions affecting the appraised assets, they should be noted in this section.

Highest and Best Use

Highest and best use as it is applicable to machinery and equipment is defined as that use of the subject machinery and equipment which may reasonably be expected to produce the greatest net return over a given period of time, that legal use which will yield the highest present value.

It is important to recognize that the highest and best use of any individual unit or total operating facility may not be its present use. It is an unsound appraisal practice to assume that the highest and best use of any item of machinery and equipment is its present use. It may be or it may not be; hence, an analysis must be made to determine the highest and best use of the assets being valued.

In general, the highest and best use of any item of machinery or production line is the use for which it was designed and built. However, this is not always true. A prime consideration in selecting any other use is the economic feasibility of converting to another use. In other words, is the prime cost of acquiring the equipment, plus the cost to convert to a different use, less than the cost of building a new facility designed and built to serve a new use?

It must be remembered that any production facility or group of machinery and equipment assets, designed and built to serve a specific function, would not be subject to the same factors or degree of functional/technological obsolescence that a converted facility would.

There are times when a total operating facility could have a greater value for offsite use than it would have in place. Thus, its highest and best use must reflect this and be so stated.

Of course, the cost of removal must be given consideration, and there are times when this cost will not cause a diminution in value; rather, it will add to the market value of the appraised assets.

If the appraisal is being made for liquidation value for offsite use, then it is evident that the subject machinery and equipment cannot have a highest and best use in place as an operating facility.

In most industrial facilities, the appraiser will find machinery and equipment that can be classed as general purpose equipment; that is, it can be used in more than one type of facility. This type of equipment should be so classed, and it should be noted that it would have a use in other manufacturing facilities.

If the item is so special that its use would be limited to one type of manufacturing operation, then its highest and best use must reflect this fact.

It must also be noted that there will be items of machinery or equipment that are so large or built into the real estate that any attempt to remove them would cause damage to the item, limit its use in any other facility, or result in the item's having only a scrap value. If this is the case, then it must be stated.

All opinions as to the highest and best use must be supported by an analysis of all factors, and the conclusion of the appraiser must be logical to the reader of the report.

Determination of Value

The purpose of the appraisal is the factor that is paramount in how this section is handled. A machinery and equipment report that has as its purpose three or four different estimates of value will naturally require more detail or a more descriptive analysis as to how the value estimates were established than one with only one estimate of value.

The various approaches to value utilized in the report should be summarized in this section. The factors of cost, depreciation, and functional, technological, and economic obsolescence, if applicable, should be discussed. It must be recognized that not all these factors will be applicable to each and every machinery and equipment report. However, those that are should be explained.

Final Estimate of Value

This portion of the report is often referred to as the *Conclusion,* or *Reconciliation.*

Regardless of the term applied, it is the portion of the report where the appraiser sets forth in a narrative form the final conclusion or estimate of the value(s) as stated in the purpose of the report.

The salient facts that were considered and analyzed in the body of the report should be summarized in this section. Here the appraiser must weigh the applicability of the value concept or concepts as they are applicable to the subject problem and with professional judgment arrive at a final conclusion that is logical and supported.

The final estimate of value is never the result of a mathematical calculation, that is, the total of all the values by the various approaches divided by the number of approaches utilized. It is rather the appraiser's professional opinion, based on the application of the various appraisal concepts and the reconciliation of these findings into a final estimate of value in accordance with the stated purpose of the report.

Certification

The appraisal report should contain certification that sets forth all the conditions that are applicable to the value estimate as reported.

There are certain conditions that would be applicable to all machinery and equipment reports. While the requirements of this section vary depending on the assignment, there are certain basic statements that form the foundation upon which each appraiser can build and tailor this section to the assignment at hand. These are included in the Code of Ethics specified by the American Society of Appraisers.

A suggested general listing of statements would include, but would not be limited to:

Statement as to a personal inspection

A statement that the appraiser has no financial interest in the items appraised (if the appraiser does, that should be stated along with the facts concerning that interest)

The fact that the fee for the appraisal is not contingent upon the value or values reported (contingent fees or a fee based on a fixed amount of value is unprofessional and unethical)

A statement that the value or values estimated are presented as the appraiser's considered opinion, based on the facts or data set forth in the report

A statement as to the confidentiality of the report

Detailed Inventory of Machinery and Equipment

This section must include a complete listing of all items appraised and identified under the first section of the report—"Identification of Property."

The details required in this section vary from one assignment to another depending on the purpose and function of the report. In general, however, this listing could be by location, that is, by building in a multibuilding facility, by floor, or department as required by the specific requirements of the assignment.

A division of the inventory by classification may be required in a million-square-foot industrial facility and not required in the valuation of the corner grocery store.

There are any number of classifications or combinations of classes of machinery and equipment, and the type or number utilized in any given report depends on the scope of the assignment and the requirements of the client.

A suggested list of standard classifications might include, but would not be limited to:

- Major machinery
- Machinery and equipment
- Dies, jigs, and fixtures
- Equipment not in use
- Cranes not in use
- Electrical power distribution system
- Process steam system
- Compressed air system
- Special built-in fixtures
- Material handling equipment
- Plant furniture and fixtures
- Office furniture
- Office machines
- Licensed vehicles

Not all these classifications will apply at all times. In fact, there will be assignments when none will be applicable and the appraiser will be required to establish the classifications that are applicable to the specific appraisal problem at hand.

Regardless of the type of machinery and equipment being valued, there are certain basic facts that must be included. The actual description of each item appraised will vary with the type of equipment being valued and the requirement of the assignment.

However, each description should provide an unqualified description of the asset being valued. This should be done in a professional manner so that the reader of the report can visualize the item and not confuse it with any other asset.

The inventory of assets appraised will, at various times and under a number of different circumstances, serve as the base for:

- The proof of loss in the event of a fire damage claim
- The proof of loss for items stolen or removed from the appraised facility

- Inclusion as part of a bill of sale
- An attachment to a collateral loan document (security attachment)
- The base for which a condemnation case is to be settled, as to what items were left on site and what items were removed

Therefore, it is important that the inventory provide an unequivocal identification of the assets being valued.

There are many ways in which items of machinery and equipment can be identified. However, within the categories utilized in any report, the following type of information, if available, should be included:

- Name of unit or type of machine or assets, including the number of units being valued
- Manufacturer's name, or any other data that will assist the reader to research this item
- Model or manufacturer's identification number (if this information is not available or is not applicable, then its size, capacity, or other data that can be used to research the item, or a comparable item if the subject is no longer a current production item, should be included)
- The manufacturer's serial number, job number, or order number
- The year manufactured, if available
- Any standard or extra attachments, type of drive, type of attachment, electrical or other power connection, such as steam, air, or hydraulics

If the asset has a special foundation pit or requires any special framing or installation, this also should be noted.

It should be noted that while these basic data are being gathered during the appraiser's inspection of the assets, it is also the time to secure any data as to the physical condition, functional utility, obsolescence factors, or any other data that will be utilized in the valuation process as it is applied to the assets being valued.

Photographs and/or Layouts

There are no defined regulations stating that it is mandatory to include photographs or layouts with each appraisal report. However, good appraisal practice suggests that this be done. It is further evidence of the effort that was put into the assignment and could be very helpful to a reader of the report who has little, if any, knowledge of the machinery and equipment being valued.

It must be recognized that it is not possible to include a photograph of each asset. Therefore, the appraiser must use professional judgment and include only those assets that are the major assets or which would present a true picture of the facility being valued.

There are instances when a layout indicating the location of the machinery and equipment would be useful. In fact, there are times when a client will request such a plan and the appraiser is instructed to make an allowance for this in the fee estimate.

Special Situations

There are a number of items that are not part of a machinery and equipment report that would normally be associated with a standard appraisal report of real estate. There are items that apply only to real estate and others that should be noted in a machinery and equipment report only under special situations. A sampling of these types of headings might include:

Description. Machinery and equipment have different descriptions: the type of machine, manufacturer, model number, serial number, size, motor horsepower, and any specific, particular, or special arrangement, such as a lease purchase agreement, lease without option to purchase, or some other form of ownership (e.g., joint).

It is also possible for an item to be leased, but the client is responsible for the cost of installation, or, in the case of fire, insurance. In all cases, if any special conditions are found, they must be noted.

Zoning. In general, machinery and equipment are not subject to local zoning regulations in the same sense as real estate. There are other regulations or government restrictions that affect machinery and equipment. Examples of these might be local fire code, health and sanitary regulations, and environmental or pure air and water laws.

A prime example of how such regulations affect machinery and equipment is in the food industry. Many facilities have equipment that was installed before the enactment of present law. Such items are legal as long as they remain in place, but they cannot be moved and reinstalled in another location. Thus, they have a value-in-place-in-use, but they may have only a limited value, if any, for offsite use.

If the appraisal was made in accordance with any special restrictions or terms imposed by the client, these must be stated. Furthermore, if the appraisal is of a fractional interest or if it is based on hypothetical conditions or a reconstruction of an asset, a statement to this effect must be included.

Conclusion

It must be recognized that each appraiser has the responsibility to remain in the possession of all the data required to support the final estimate of value. A copy of the finished report submitted to the client must also be kept in the appraiser's file.

Finally, each appraiser or appraisal company has the professional responsibility to maintain the confidentiality of the findings and report. The appraisal report is the client's property; the appraiser is only the keeper of the data contained in it. Therefore, no one has the right to know the data, analysis, or conclusion of value stated in that report without the permission of the client, that is, the one who authorized and paid for that report.

There are times when, owing to the process of the law, an appraiser must give that information. However, in general, the appraiser must not violate the confidential nature of the client-appraiser relationship.

14
Ethics

John Alico, P.E., FASA
*President, Alico Engineers and Appraisers
Birmingham, Michigan*

Each of the appraisal societies which educate, test, and certify qualified appraisers has a Code of Ethics. This discussion is based on the Code of Ethics promulgated by the American Society of Appraisers, the only multidisciplinary appraisal organization in the United States. It follows the outline of the ethics course presented on audiovideo tape which is made available to members through their respective chapters.

To begin with, what is ethics? Possibly the best way to answer this is to explain that a Code of Ethics is a system by which professional behavior is to be governed. Further, the system of ethics that we use in appraisal practice is designed not only to protect our clients but also to enhance the professional well being of other appraisers.

Why do we require a system of ethics? Perhaps the simplest answer is that it tells us how to act, what actions are right or wrong, good or bad. This could mean that ethics may be defined as a science or practice which deals with conduct insofar as it is considered right or wrong, good or bad.

The appraiser's prime concern is the determination of value. Judgment is an important factor in arriving at this determination. Basically, we recognize that there are right or wrong ways of doing things. For instance, there is a right way and a wrong way to build a bridge, a right way and a wrong way to assist in a surgical operation; there are also right ways and wrong ways of doing things in the larger relations of life, in the business of life itself, and it is with these things that ethics is concerned.

Let us look on the other side of the coin. Let us imagine that the need for ethics is not important because human beings already know, without any specific training or education, what is a good thing to do and what is a bad thing to do. However, whether we are so gifted is a matter of judgment. The obligation to understand ethical values is almost the first law of life, and the place where it is in the highest demand is in the world of human and moral relations.

Ethics is not a new idea, it has existed as long as human beings began to think. Plato, Socrates, Aristotle, and Aristophanes are some of the early philosophers who defined ethics and expressed their opinions on what is good and what is bad.

Appraisers Are Obligated to Attain Competency and to Practice Ethically

One of the differences between a professional and a nonprofessional is the Code of Ethics to which the professional must subscribe. This is established by a governing body or professional society of which the recipient is a member. By becoming a member of an established, accrediting organization, the professional places himself or herself under the jurisdiction of that governing body and is subject to disciplinary action in the event that the society's Code of Ethics is not followed.

When accepting an assignment, the accredited appraiser places himself or herself within the Code of Ethics of the society or organization that granted the designation. It is essential that the purpose of the appraisal be defined so that the appraiser can select the appraisal method and analysis which must be employed to give a logical opinion of the value which is to be found. After selecting the method to be utilized, it is the appraiser's obligation to determine the appropriate and applicable numerical results with as high a degree of accuracy as the particular objectives of the appraisal necessitate.

For example, we assume that the appraiser is determining the fair market value of an item of machinery and equipment. The purpose of the appraisal is stated to be the determination of the fair market value. The appraiser should define fair market value so that whoever reads the appraisal report will know that the value is based on a set of specific conditions. For example, the fair market value may apply to all the equipment to be sold as a unit for use in place.

In this instance, the appraiser should say that the fair market value is based on this condition. On the other hand, it may be that the fair market value is to be found for the equipment for purchase, in whole or in part, for offsite use. In this case, the appraiser should know and understand that the equipment has to be moved and that moving costs are generated by the size and weight of the item and the necessity to supply new foundations, if any should be required.

This kind of expertise requires that the appraiser must be competent in his or her field. Competency is obtained by education, training, study, practice, and experience.

The appraiser must also recognize the problems that might be generated by moving a piece of equipment. Sometimes a unit, such as a large boiler, is installed, and the building is built around it. This kind of unit may not be moved readily and without great expense. The fair market value for offsite use would be considerably lower than the fair market value for use-in-place. This is the kind of consideration which the appraiser must give to a problem. Therefore, the appraiser must have information, expertise, and knowledge, which is not normally possessed by a layperson, and this is part of the difference between the professional and a person who cannot make this claim.

The Appraiser's Fiduciary Relationship to Third Parties

The client, recognizing that the appraiser has expertise which is specialized, and which the client himself does not readily have, must depend on the appraiser to accomplish the objectives of the project. There is no caveat emptor principle involved in a relationship between a professional appraiser and a client. On occasion, it may be that the client will turn the appraisal report over to a third party. It is essential that the appraiser recognize that the third party or parties have as much right to rely on the validity and objectivity of the appraisal as does the client. This may be the case where the client submits the appraisal to a banker for review as collateral evidence.

The Appraiser's Fiduciary Relationship to the Public

The fiduciary relationship to the public is the same as the appraiser's fiduciary relationship with third parties. This would apply to assignments involving depositors in a financial institution making loans, to taxpayers in a school district whose board is acquiring a new school, to taxpayers represented by government agencies who are acquiring property under an eminent domain proceeding, and to publicly displayed values of real or personal property that is offered for sale by a government agency.

The Appraiser's Obligations to the Client

In some cases, the appraiser turns over the report to a reviewer designated by the client. This is a proper procedure, and the appraiser is bound by it inasmuch as this is a condition set forth by the client. On the other hand, the appraiser must keep the appraisal a confidential matter, which involves only the client. The appraiser cannot reveal the property being appraised, the identity of his client, or the results of the appraisal. This is all a matter of confidential information which is to be delivered only to the client.

It is not proper for the appraiser to reveal to anyone other than the client the amount of the valuation of the property. Sometimes the client will give the appraiser approval to release the information to a specific person or body, in which case the appraiser should obtain written notice to this effect in order to protect himself or herself and to ensure that he or she is handling the matter in an ethical manner. An appraiser cannot use an appraisal report made for a client as evidence of professional qualifications to an appraisal or professional society unless the client has given permission for this type of review.

When an appraiser accepts an assignment, it is with the understanding that he or she is qualified to do the work for which he or she is engaged and that the appraiser's field of expertise does not lie outside the appraisal problem. It is possible for an appraiser to accept the assignment, provided that he or she acquaints the client with the limitations of his or her own qualification or associates himself or herself with another appraiser or appraisers who possess the required qualification.

The Appraiser's Obligation Relative to Giving Testimony

In the event that the appraiser testifies in court or gives testimony in a deposition, it is necessary to present the data analysis and value without bias regardless of the effect of such unbiased presentation on the client's case. An appraiser must not suppress any facts or opinions which are adverse to the case the client is trying to establish or to overemphasize any facts or opinions which are favorable to the client's case. In other words, the appraiser cannot ethically become an advocate. The appraiser may not properly serve more than one client with respect to the same property, or the same legal action, without the consent of both parties to act in this capacity.

The Appraiser's Obligation to Other Appraisers

The appraiser has obligations to other appraisers. For example, an appraiser should protect the professional reputation of all appraisers who subscribe to and practice in accordance with the principles of appraisal practice of an accrediting organization such as the American Society of Appraisers. It is unethical for an appraiser to injure, or attempt to injure, by false or malicious statement or innuendo, the professional reputation of any other appraiser.

Unethical Competitive Conduct

An appraiser should not reduce a fee which he or she has already quoted to a client to obtain an appraisal contract and thus supplant another appraiser after the latest quotation has been made known. As a matter of fact, it is unethical for an appraiser to supplant, or attempt to supplant, another appraiser after the latter has been engaged to perform a specified appraisal service. It is the obligation of a member of the society who has knowledge of an unethical act on the part of another member, to report that act to the society. Further, it is the appraiser's obligation to cooperate with the society in all matters, including investigation, censure, discipline, or dismissal of members who are charged with violations of the society's Code of Ethics. An appraiser, acting ethically, cannot indulge in self-laudatory advertising or solicitation of appraisal engagements using unwarranted, inaccurate, or misleading claims or promises. These are acts which are considered detrimental to the establishment and maintenance of public confidence in the results of appraisal work. Advocacy, of course, is another aspect of the same problem. Misrepresentation as to qualifications and membership status are also considered unethical practices and are not to be engaged in by appraisers.

Acceptance of Contingency Fees Is Unethical

If an appraiser were to accept an engagement for which the amount of compensation is contingent upon the amount of an award in a property settlement or a court action where the appraiser's services are employed, or is contingent upon the amount of a tax reduction obtained by a client where the appraiser's services are employed, or is contingent upon the amount of a tax reduction obtained by a client where the appraiser's services are used, or is contingent upon the consummation of the sale or financing of a property in connection with which the appraiser's services are utilized, or is contingent upon the appraiser's reaching any finding or conclusion specified by the client, then anyone considering using the results of the appraiser's undertaking might well suspect that these results were biased and self-serving and, therefore, invalid. Such suspicion would militate against the establishment and maintenance of trust and confidence in the results of appraisal work generally. Therefore, the society considers that the contracting for, or acceptance of, any such contingent fee is unethical and unprofessional.

As a corollary to the above principle relative to contingency fees, the society considers that it is unethical and unprofessional for an appraiser (1) to contract for or accept compensation for appraisal services in the form of a commission, rebate, division of brokerage commissions, or any similar forms and (2) to receive or pay finders' or referral fees.

In the matter of percentage fees, the society takes the position that it is unprofessional and unethical for the appraiser to contract to do work for a fixed percentage of the amount of value, or of the estimated cost (as the case may be) which the appraiser determines at the conclusion of the work.

Disinterested Appraisals

Anyone using an appraisal made by an appraiser who has an interest or a contemplated future interest in the property appraised might well suspect that the report was biased and self-serving and, therefore, that the findings were invalid. Such suspicion tends to break down trust and confidence in the results of appraisal work in general. It is, therefore, necessary and proper that the appraiser be wholly disinterested in the property being appraised.

Interests which an appraiser should avoid in a property being appraised include ownership of the subject property; acting, or having some expectation of acting, as agent in the purchase, sale, or financing of the subject property; and managing, or having some expectation of managing, the subject property. Such interests are particularly apt to exist if the appraiser, while engaged in professional appraisal practice, is also engaged in a related retail business (real estate, jewelry, furs, antiques, fine arts, etc.).

The society professes that, subject to the provision for disclosure given in the following paragraph, it is unethical and unprofessional for an appraiser to accept an assignment to appraise a property in which he or she has an interest or a contemplated future interest.

However, if a prospective client, after full disclosure by the appraiser of his or her present or contemplated future interest in the subject property, still desires to have the appraiser do the work, the latter may properly accept the engagement provided he or she discloses the nature and extent of interest in the appraisal report.

Responsibility Connected with Signatures to Appraisal Reports

The user of an appraisal report, before placing reliance on its conclusions, is entitled to assume that the party signing the report is responsible for the findings, either because he or she did the work or because the work was done under his or her supervision.

In cases where two or more appraisers are employed to prepare a joint report, the user is entitled to assume that if all of them sign it, they are jointly and severally responsible for the validity of all the findings therein, and if all do not sign, the client has a right to know what the dissenting opinions are.

In cases where two or more appraisers have been engaged by a single client to make independent appraisals of the same property, the client has the right to expect that he or she will receive opinions which have been reached independently, and that they may be used as checks against each other and/or have evidence of the range within which the numerical results lie.

To implement these principles, the society says that it is unethical (1) to misrepresent who made an appraisal by appending the signature of any person who neither did the work nor had the work done under his or her supervision; (2) in the case of a joint report to omit any signatures or any dissenting opinions; (3) in case two or more appraisers have collaborated in an appraisal undertaking, for them, or any of them, to issue separate appraisal reports; and (4) in case two or more appraisers have been engaged by a single client to make independent appraisals of the same property, for them to collaborate or consult with one another or make use of each other's findings or figures.

An appraisal firm or corporation may properly use a corporate signature with the signature of a responsible officer. But the person who actually did the appraisal for the corporation must sign the corporate appraisal report, or the report must acknowledge the person who actually made the appraisal.

Unconsidered Opinions and Preliminary Reports

If an appraiser gives an opinion as to the value, earning power, or cost estimate of a property without having ascertained and weighed all the pertinent facts, such opinion, except by an extraordinary coincidence, will be inaccurate. The giving of such offhand opinions tends to belittle the importance of inspection, investigation, and analysis in appraisal procedure and lessens the confidence with which the results of good appraisal practice are received, and therefore the society professes that the giving of hasty and unconsidered opinions is unprofessional.

If an appraiser makes a preliminary report without including a statement to the effect that it is preliminary and the figures given are subject to refinement or change when the final report is completed, there is the possibility that some user of the report, being under the impression that it is a final and completed report, will accord the figures a degree of accuracy and reliability they do not possess. The results of such misplaced confidence could be damaging to the reputation of professional appraisers in general as well as to that of the appraiser concerned. To obviate this possibility, the society considers it to be unprofessional appraisal practice to omit a proper limiting and qualifying statement in a preliminary report.

Advertising and Solicitation

It is not unethical to advertise the availability of appraisal services. It is unethical to use any inaccurate, misleading, false, or deceptive claim, promise, or representation in connection with any advertisement. These unethical practices are considered by the society in the results of appraisal work. The society considers that such practices on the part of an appraiser constitute unethical and unprofessional conduct. It would be unethical to do the following:

1. Misrepresent in any way one's connection or affiliation with the American Society of Appraisers or any other organization
2. Misrepresent one's background, education, training, or expertise
3. Misrepresent services available or an appraiser's prior or current service to any client, or identify any client without the express written permission of such client to be identified in advertising
4. Represent, guarantee, or imply that a particular valuation or estimate of value or result of an engagement will be tailored or adjusted to any particular use or conclusion other than that an appraisal will be based on an honest and accurate adherence to the principles of appraisal practice as published by the American Society of Appraisers.

Misuse of Membership Designation

The constitution and bylaws of the society establish three professional grades of membership, namely, member, senior member, and fellow. (An affiliate or candidate does not hold a professional grade of membership in the society.) Members may use the designation "Member of the American Society of Appraisers." Only senior members may use the designation "A.S.A." Only fellows may use the designation "F.A.S.A." The society considers it to be unethical for a member to claim or imply the holding of a higher degree of membership than he or she has attained.

Summary

Disciplinary action against the members of the society is taken in the event of violations of specific provisions of the society's constitution and bylaws or of its principles of appraisal practice and the Code of Ethics incorporated in it. Such actions are under the jurisdiction of the international president, the international ethics committee, and the board of governors. Violations may fall under four categories:

1. Deviations from good appraisal practice
2. Failure to fulfill obligations and responsibilities
3. Unprofessional conduct
4. Unethical conduct

After investigation, the society may take action in the form of suggestion, censure, suspension, or expulsion, in which event the member will be required to surrender his or her certificate, membership pin, and other evidences of membership and to desist from all reference to such membership.

The designated members, affiliates, and candidates making up the American Society of Appraisers are governed by the above code. Moreover, there is the implied obligation on the part of the entire membership to uphold and maintain the integrity which the general public has come to expect from the members of a professional organization. There is a good way to do things in the appraisal profession and the society has taken a stand against the bad ways. By doing so, a right way and a wrong way of making an appraisal have been defined because we aspire to the highest ideals in accomplishment and recognize that professional performance requires strict and steadfast fidelity to ethical behavior. All members, including affiliates and candidates, must fully ascribe to the Code of Ethics as set forth by the American Society of Appraisers.

Index

Actual cash value, 35, 154, 157
Advertising, 198–199
Aesthetic values, 3
Agabian, Merritt, 39–48
Age/life technique, 90–94
  overall depreciation and, 107–109
Age of unit, 112
Aircraft Blue Book, 53
Alico, John, 59–78, 191–200
Allocation of purchase price, 32
Alternative uses, 143–144
American Society of Appraisers (ASA), 61, 80, 167
  ethics and (*see* Ethics)
  membership designations, 199
Amortized assets, 12
Anticipated liability, 149–151
Appraisal of Equipment, Inventory, and Supplies, The, 78n.
Appraisal of Machinery and Equipment, The (ASA), 42n., 44n., 47n., 60n.
Appraisal of Real Estate, The (AIREA), 9n., 10n., 104
Appraisal Principles and Procedures (Babcock), 6, 14n., 15n., 60n., 61n.
Appraisal procedure, 61–64
  cost method, 62–64
  depreciation and, 63–64
  (*See also* Depreciation)
  inspection, 61–62
  purposes of equipment and, 62
  steps in, 176
Appraisal report, 175–189
  assembly of, 144–145
  certification, 185
  conclusion, 189
  or reconciliation, 184–185

Appraisal report (*Cont.*):
  described, 175
  description of machinery and equipment, 182–183
  detailed inventory, 185–187
  determination of value, 184
  ethics and, 175–176, 185
  final estimate of value, 184–185
  general statement for liquidation studies, 150–151
  highest and best use, 183–184
  identification of property, 178
  letter of transmittal, 144–145, 176–177
  liability and, 149–151
  minimum requirements for, 176
  photographs and/or layouts, 187–188
  preliminary reports, 198
  purpose of appraisal, 178–180
  (*See also* Purposes of appraisal)
  real estate and, 188
  responsibility connected with signatures to, 197–198
  special situations, 188–189
  statements in, 185
  zoning and, 188
Appraisal Terminology and Handbook (AIREA), 66n.
Appraisals, purposes of (*see* Purposes of Appraisals)
Assumptions and Limiting Conditions, statement of, 18
Auction sales, 132
  monitoring of, 135, 139
  (*See also* Liquidation value concepts)
Auctioneers, 130
Averaging, 132

Babcock, Henry, *Appraisal Principles and Procedures*, 6, 14n., 15n., 60n., 61n.
Bankruptcy, 32–33, 130
Barrow, Kal, 153–160
Betterment, 103–104
Black, James H., 126
Black's Law Dictionary (Nolan), 12n.
Blue Book-Auto, 53
Blue Book of Current Market Prices of Used Heavy Construction Equipment, 53
Bodily values, 2
Book values, 73
Boyce, Bryl N., *Real Estate Appraisal Terminology*, 59n., 61n., 81n.
Brightman, Edgar Sheffield, *Preface to Philosophy*, 2–4
Building Construction Cost Data, 53, 164
Business enterprise, 119–121
Business valuation, 33

Canonne, Jean, 4–5, 7
Capital cost, 83
Catalogs, manufacturer's, 53
Causes and effects, 140–144
  (*See also* Correlation)
Character values, 3
Chattle fixture, 11
Chilton, C. H., 40
Classification of property, 9–16
  fixtures, 11
  intangible property, 12–13
  personal property, 10–11, 13–15
  real estate, 9–10
  special cases, 11–12
  state laws and, 13–15
  summary, 15–16
  for valuation, 6–7
Clients:
  documents of, 142
  obligations to, 194
Code of Ethics (*see* Ethics)
Coinsurance, 160
Commercial bankers, 129–130
Communication, 137
Comparable match, 113
Computer Hardware & Software, 53
Computer Price Guide, 54
Condemnation, 33
Condition of unit, 112
Connolly, John J., III, 9–16
Contingency fees, 196
Continued-use premise, 79–80

Continued-use premise (*Cont.*):
  determining value for, 114–118
Correlation, 139–144
  alternative uses, 143–144
  causes and effects and, 140–144
  client documents and, 142
  conclusion derived from, 142–144
  ease of removal, 144
  industry location and economics, 141
  intuitive value indicators, 143
  negative marketing aspects, 143
  physical appearance, 140–141
  positive marketing aspects, 143
  psychological effects, 141–142
  special considerations, 143–144
  total draw, 140
Cost:
  historical, 60–61
  original, 60
  upper limit of, 83n.
  value versus, 48, 60
Cost and Optimization Engineering (Jelen and Black), 126
Cost approach to fair market value, 81–110
  defined, 82
  depreciation and, 86–109
    economic obsolescence, 104–107
    functional obsolescence (*see* Functional obsolescence)
    overall, 107–108
    physical deterioration (*see* Physical deterioration)
    reference table for, 63–64
    sequence of, 108–109
  determination of current cost, 82–86
    cost of replacement and, 82–86
    cost of reproduction and, 82–83, 85–86
    proper level of current cost, 84–85
    priniciple of substitution and, 81–84
  starting point in, 34
  steps in, 62–63, 82
  strengths and weaknesses of, 122–123
  summary, 109–110
Cost data procedures, 44
  cost sources (*see* Sources of pricing and reference material)
  other methods of determining costs, 46
  trending cost data, 44–45
Cost Engineers Notebook (AACE), 126
Cost index, effective age and, 90–92
Cost studies, 33–34

Index   **203**

Cost-to-capacity relationship, 125–127
  inutility penalty and, 104–107
Cost-to-cure, 62
Curable physical deterioration, 94–96

Davis, Nobel L., 45n.
Declining-balance depreciation, 73–74
Definition in reports, 145
Depreciation, 59–78
  accounting methods used to determine, 73–76
    declining balance, 73–74
    interest theories, 76
    noninterest procedures, 73–75
    present worth, 76
    sinking fund, 76
    straight line, 73
    sum-of-the-years'-digits, 74–75
  appraisal procedure and, 61–64
    cost method, 62–63
    inspection, 61–62
    purpose of equipment, 62
  cost and, 60–61
  cost-to-cure and, 62
  defined, 64
  definitions of condition, 63, 67–68
  effects of condition on, 66–67
  fair condition, 63, 68
  good condition, 63, 68
  historical cost and, 60–61
  machines and, 59
  market data approach as applied to, 71–73
  new condition, 63
  obsolescence and, 69–71
    economic, 70–71, 104–107
    functional (*see* Functional obsolescence)
    technological, 69
  original cost and, 60
  physical deterioration (*see* Physical deterioration)
  poor condition, 63, 68
  reference table, 63–64
  reserve to provide for replacement and, 65
  scrap, 62, 63, 68
  of set-up costs for used machines, 115
  summary and conclusions, 77–78
  usable condition, 63
  useful life expectancy, 64–66

Depreciation (*Cont.*):
  value and, 59–61
  very good condition, 63, 68
Depreciation (IRS), 65
Desk-top studies, 147
Deterioration, 66
Dictionary of Scientific and Technological Terms, 54
Direct dollar measurement of physical deterioration, 94–96
Direct match, 113
Directory of Industry Data Sources, 54
Disinterested appraisals, 196–197
Dissolutions, 34

Ease of removal, 144
Economic life, 62–63
Economic obsolescence, 70–71, 104–107
  causes of, 104
  defined, 104
  income approach and, 119–121
  inutility penalty and, 104–107
Economic supportability, 121–122
Economic values, 2
Effective age of equipment, 90–92
Energy Cost Reference Book, 54
Engineering Valuation and Depreciation (Marston et al.), 64n., 66n., 67n., 76n., 94, 108, 153n.
Equipment Directory of Audio and Visual Computer and Video Products, 54
Erection or assembly, 46, 47
Estate planning, 34
Ethics, 191–200
  advertising and solicitation, 198–199
  appraisal report and, 175–176, 185
  contingency fees and, 196
  defined, 191
  disinterested appraisals, 196–197
  fiduciary relationships: to the public, 193–194
    to third parties, 193
  misuse of membership designation, 199
  obligations: to attain competency and practice ethically, 192–193
    to the client, 194
    to give testimony, 194–195
    to other appraisers, 195
  preliminary reports, 198
  responsibility connected with signatures to appraisal reports, 197–198
  summary, 199–200

Ethics (*Cont.*):
  unconsidered opinions, 198
  unethical competitive conduct, 195–196
Excess capital costs, 83
  functional obsolescence from, 97–98
Excess operating costs, functional obsolescence from (*see* Operating obsolescence)

Fair condition, 63, 68
Fair market value, 79–127
  appendix, 125–127
  basic concepts of, 79–81
    continued-use premise and, 79–80
    fair market value-in-place and, 80–81
  conclusion, 125
  cost approach to (*see* Cost approach to fair market value)
  income approach to, 119–121
  liquidation and, 131
  market approach to (*see* Market approach to fair market value)
  need to qualify appraisals and, 121–122
  for off-site use, 179–180
  review, 122–125
  value-in-use/value-in-exchange and, 170–172
Features (accessories), 112
Fiduciary relationships:
  to the public, 193–194
  to third parties, 193
Final analysis for application to comparable indicators, 144
Fire insurance, 153
Fixtures, 11
Food Processors Guide, 54
Forced sale, 130–131
  quasi, 131
  (*See also* Liquidation value concepts)
Formula/ratio method of measuring physical deterioration, 89–94
  defined, 89
  determination: of effective age, 90–92
    of physical condition, 92–93
  use of, 89
Fractional appraisal, 80, 81
Franklin, Benjamin, 153
Free publications, 52–53
Freight charge, 46, 47
Functional obsolescence, 69–70, 97–104
  defined, 97

Functional obsolescence (*Cont.*):
  examples of, 100–104
  from excess capital costs, 97–98
  from excess operating costs, 98–100

Garber, Henry A., 44*n*.
Good condition, 63, 68
Goodwill, 12
Graham, David M., 29–38, 161–165
Green Guide for Lift Trucks, 54

Hempstead, Jean C., 64*n*., 94
Highest and best use, 183–184
Historical cost, 60–61
Hypothetical situations, 147–148

Iannacito, Alan C., 17–28
Identification of machinery and equipment, 17–28
  appendix, 27
  leased equipment, 17–18
  macroidentification, 18–23, 27
  microidentification, 23–26
  recommended readings, 28
IMN Auction Report, 54
Impersonal value, 7
Income approach to fair market value, 119–121
  amount supportable and, 120–121
  business enterprise and, 120
  strengths and weaknesses of, 124–125
Incorporation procedure, 34–35
Indirect charges, 46, 47
Industrial Machinery News, 54, 71*n*.
Industrial Real Estate (Kinnard), 61*n*., 76*n*.
Industry location and economics, 141
Inspection of property, 61–62
Installation charges, 46, 47
Instrumental values, 2
Insurable value, 153–160
  actual cash value and, 35, 154, 157
  coinsurance, 160
  conclusion, 160
  defined, 154
  early development, 153–154
  extent of detail in appraisal report, 158
  fire insurance and, 153–154
  insurable replacement cost, 154
    procedure for determining, 158–159
  insurable value depreciated, 154
    procedure for determining, 159

Insurable value (*Cont.*):
  loss settlement and, 35
  methodology, 158
  as purpose of appraisal, 35
  requirements for appraisal for, 154, 157
  "standard 165 lines" of, 154–157
  written on a replacement form, 33–34
Intangible assets, 12–13
Intangibles, 148–149
Intellectual values, 3
Interest procedures to determine depreciation, 76
Intermediate term lenders, 129–130
Internal financial considerations, 36
Interpretation page in reports, 145
Intrinsic values:
  higher, 3–4
  lower, 2–3
Intuitive value indicators, 143
Inutility penalty, 104–107
Investment property, 6, 7
Investor buyers, 173

Jelen, Frederic C., 126
Joint venture, 36

Kinnard, William N., Jr., *Industrial Real Estate*, 61n., 76n.

Last Bid, The, 55
Leased equipment, 17–18
Letter of transmittal, 144–145, 176–177
  sample, 177
Liability, 149–151
Life expectancy, useful, 64–66
Liquidation, 130–132
  averaging and, 132
  forced sale and, 130–131
Liquidation value concepts, 129–151
  anticipated liability, 149–151
  appraisal report and, 180–182
  as auction sale, 132
  auctioneers and, 130
  bankruptcy and, 32–33, 130
  correlation and (*see* Correlation)
  definitions, 130–134
    liquidation value, 132, 180
    liquidation value-in-place, 133–134, 180
    orderly liquidation value, 132–133, 180

Liquidation value concepts (*Cont.*):
  ease of removal and, 144
  equipment or plant purchasing and, 130
  experience required for, 134–137
    areas of, 134
    auction monitoring, 135
    communication, 137
    understanding the industry, 135–137
  fair market value and, 131
  forced sale and, 130–131
    quasi, 131
  methods of valuation used for, 137
    listing, 138
    research library, 138–139
    nonstandard studies (*see* Nonstandard studies)
  overview, 151
  secured lenders and, 129–130
  uses of, 129–130
  value-in-use/value-in-exchange and, 170–171
Liquidation value-in-place, 133–134, 180
  anticipated liability and, 150
  value-in-use and, 170–171
Loans, appraisal for, 35
Location of asset, 112
Locator of Used Machinery and Equipment, 71n.

Machine, definition of, 59
"Machinery and Equipment Trends—How Are They Used?," 45n.
Macroidentification, 18–23, 27
Management considerations, 36
Manufacturer of asset, 112
Manufacturer's catalogs, 53
Market approach to fair market value, 110–119
  applied to total properties, 111
  comparable match, 113–114
  defined, 110
  determining value for continued use, 114–118
  direct match, 113
  elements of comparability, 111–113
  logic behind, 110
  percent of cost, 114
  steps in, 71, 111
  strengths and weaknesses of, 123–124
  summary, 119
Market conditions, 112, 118

Market data approach as applied to depreciation, 71–73
Market value-in-place, 71
   as purpose of appraisal, 179–181
Marketable property, 6–7
Marketing aspects, 143
Marshall Valuation Service, 54
Marston, Anson, 64n., 66n., 67n., 76n., 94
Martin, Kenneth A., 1–8
Merger, 36
Microidentification, 23–26
Miles, Leslie H., Jr., 129–151, 167–174
Modern Cost Engineering Methods and Data, 54
Motivation, 112

National Construction Estimator, 164
Natural values, 2
Negative marketing aspects, 143
New condition, 63
Nolan, Joseph R., *Black's Law Dictionary*, 12n.
Nomda's Blue Book Industrial Guide to Market Prices &Manufacturers' Trade-In Schedules, 54
Noninterest procedures to determine depreciation, 73–76
Noninvestment property, 6–7
Nonstandard studies, 145–149
   desk-top studies, 147
   hypothetical situations, 147–148
   intangibles, 148–149
   walk-through studies, 146–147
   word "appraisal" and, 145–146
North American Conference of Appraisal Organizations, 176

Obsolescence, 69–71
   additive nature of, 70–71
   economic, 70–71
   functional, 69–70
   technological, 69
   value-in-use/value-in-exchange and, 172
Official Guide—Tractors and Farm Equipment, 54
"On the Prehistoric Origins of Ownership," 4, 7n.
Operating obsolescence, 98–100
   areas to be investigated, 99
   betterment, 103–104
   capitalization rate and, 98–99, 102–103

Operating obsolescence (*Cont.*):
   defined, 98
   estimate of penalty for continued use, 98
   examples of, 100–102
   occurrence of, 99–100
   variable operating costs and, 103
Orderly liquidation value, 132–133, 139, 180
   anticipated liability and, 150
   ease of removal and, 144
Original cost, 60
Ownership and value, 4–5

Partnership formation, 36
Percent of cost technique, 114
Personal property:
   defined, 10
   division of coverage and, 13
   effect of state laws on classification of, 13–15
   examples of, 10–11
   fixtures as, 11
   value of, 7
Physical appearance, 140–141
Physical condition, 67, 69
   determination of, 92–93
   as element of comparability, 112
Physical deterioration, 86–97
   causes of, 86–87
   as comparison between subject and new asset, 88
   curable, 94–96
   defined, 86
   effective age and, 90–92
   maintenance and, 87–88
   measuring, 87–96
      direct dollar, 94–96
      formula/ratio, 89–94
      observation, 88–89
      physical condition and, 92–93
   sources of information for, 96–97
   use of an asset and, 87
Poor condition, 63, 68
Positive marketing aspects, 143
Pratt, Shannon P., *Valuing Small Business and Professional Practices*, 104
Preface to Philosophy (Brightman), 2–4
Preliminary reports, 198
Present worth, 98–99
   depreciation and, 76

Index

Price of asset, 112
Price guides, published, 53–55
Principle of substitution, 81–82
   cost of replacement and, 83–84
Principles of Appraisal Practice and Code of Ethics, 175–176
   (*See also* Ethics)
Principles of Appraisal Practice and Code of Ethics (ASA), 61n., 80
Project and Cost Engineers Handbook (AACE), 126
Property, 5–7
   bundle of rights for, 5
   classification of (*see* Classification of property)
   concept of, 5–6
   inspection of, 61–62
   intangible, 12–13
   investment of, 6, 7
   noninvestment, 6, 7
   personal (*see* Personal property)
   real (*see* Real property)
Psychological effects, 141–142
Purchase price, allocation of, 32
Purchase/sale of asset, 36–37
Purposes of appraisals, 29–38
   allocation of purchase price, 32
   appraisal report and, 178–180
   bankruptcy, 32–33
   business valuation, 33
   condemnation, 33
   cost studies, 33–34
   dissolutions, 34
   estate planning, 34
   incorporation, 34–35
   insurance, 35
   insurance loss settlement, 35
   loan, 35
   management considerations, 36
   market value-in-place, 179
   market value for off-site use, 179–180
   merger, 36
   multiple, 31
   partnership formation, 36
   purchase/sale, 36–37
   statement of, 29, 30, 38
   stock issue, 37
   surplus disposition, 37
   taxation, 37
   third-party requirements and, 30–31
   value concepts and, 31–38

Purposes of appraisals (*Cont.*):
   value-in-use, 179

Quality of asset, 112
Quantity of asset, 112–113
Quasi forced sale, 131

Real estate, 9–10
   appraisal report and, 188
Real Estate Appraisal Terminology (AIREA), 11n., 15n.
Real Estate Appraisal Terminology (Boyce), 59n., 61n., 81n.
Real property, 9–10
   division of coverage and, 12–13
   fixtures and, 11
   special cases, 11–12
Recapitulation page, 145
Recreational values, 2
Religious values, 3–4
Removal, ease of, 144
Replacement cost new, 33–34, 39–48
   components of, 46–48
   conclusions, 48
   cost versus value, 48
   cost approach and, 81–84
   cost data procedures, 44
   defined, 39–40, 62
   determining, 41
   information required, 42–43
   installed, 39, 64
   may be greater than reproduction cost new, 85–86
   other methods for determining costs, 46
   reproduction cost new and, 40–42
   sources of, 40
   special purpose machines, 41–42
   trending cost data, 44–45
   value-in-use/value-in-exchange and, 172
Reports (*see* Appraisal report)
Reproduction cost new:
   defined, 40–41, 62
   as first step in cost approach, 62, 82–83
   replacement cost new may be greater than, 85–86
   special purpose machines and, 41–42
Research library, 49–50, 138–139
   (*See also* Sources of pricing and reference material)
Rice, Paul, 49–58

Richardson Engineering Services, Inc.—Process Plant Construction Estimating Standards, 54
Rigging costs, 46, 47
Rights of ownership, 5–6
Roberts, Thomas L., 47n., 60n.

Sale, type of, 113
Sales comparison approach (*see* Market approach to fair market value)
Schropp, Thomas L., 42n.
Scrap/salvage, 62, 63, 161–165
　analysis, 163–164
　defined, 68, 161–162
　determining value, 162–163
　net value, 162
　sources of value information, 164–165
Secured lenders, 129–130
Serial Number Book: Reference Book for Metalworking Machinery, 54
Service life, 65
Sinclair, George D., 42n., 175–189
Sinking-fund depreciation, 76
Six-tenths factor, 40
Size/type of asset, 113
Smyth v. *Ames*, 153
Social values, 3
Solicitation, 198–199
Sources of pricing and reference material, 49–58
　approaches to, 51–52
　basic, 50–51
　free publications, 52–53
　importance of, 49–50
　manufacturer's catalogs, 53
　metalworking machinery manufacturers, 55–57
　price guides, 53–55
　recommended readings, 57–58
　sharing of, 51–52
Special considerations in reports, 145
Special purpose machines, 41–42
Statement of appraisal purpose, 29, 30
　examples of, 38
Statement of limiting conditions, 145
Stock issue, 34–35, 37
Straight-line depreciation, 73, 93–94
Sum-of-the-years' digits depreciations, 74–75
Support data, 49–50

Support data (*Cont.*):
　sources of (*see* Sources of pricing and reference material)
Surplus disposition, 37
Surplus Record-Index of Available Capital Equipment, 55
Svoboda, Robert S., 79–127

Tax Reform Act of 1984, 181–182
Taxation, 47
　ad valorem, 37
　depreciation and, 65
　　(*See also* Depreciation)
　on donated property, 181–182
　gift, estate, 37
Technological obsolescence, 69
　defined, 97
Teitell, Conrad, *Taxwise Giving*, 181n.
Testimony, legal, 149
　appraisers obligations relative to, 194–195
Third-party requirements for an appraisal, 30–31
Thomas' Register Catalog File, 55
Thomas' Register of American Manufacturers, 55
Time of sale, 113
Total draw, 140
Trade fixtures, 11
Trending cost data, 44–45
　effective age and, 90–92
Truck Blue Book, The, 55
Type of sale, 113

Unconsidered opinions, 198
Understanding the industry, 135–137
Usable condition, 63
Used Equipment Directory, 55
Useful life expectancy, 64–66
User buyers, 173

Valuation (ASA), 176n.
Value(s), 1–8
　aesthetic, 3
　appraisal purposes and, 31–38
　bodily, 2
　character, 3
　cost approach to, 34
　cost versus, 48, 60
　defined, 59

Value(s) (*Cont.*):
  depreciation theory and, 59–61
  economic, 2
  higher intrinsic, 3–4
  impersonal, 7
  intellectual, 3
  lower intrinsic, 2–3
  of machinery and equipment, 7–8
  natural, 2
  ownership's relation to, 4–5
  property and, 5–7
  purely instrumental, 2
  recreational, 2
  religious, 3–4
  social, 3
  summary, 8
  upper limit of, 39, 61, 83
  work, 2–3
Value-in-place, 71
  as purpose of appraisal, 179–181
Value-in-use/value-in-exchange, 167–174

Value-in-use/value-in-exchange (*Cont.*):
  confusion between, 169–170
  defined, 167–169
  fair market value and, 170–172
  liquidation concepts and, 170–171
  as purpose of appraisal, 179–181
  replacement concepts and, 172
  summary, 174
  when they are equal, 173–174
  when they are not comparable, 172–173
Value opinions (*see* Nonstandard studies)
Valuing Small Business and Professionaal Practices (Pratt), 104
Variable operating costs, 103
Very good condition, 63, 68

Walk-through studies, 146–147
Winfrey, Robley, 64*n*., 94
Work values, 2–3

Zoning, 188

About the Authors

THE AMERICAN SOCIETY OF APPRAISERS, established in 1952, is an international nonprofit organization that promotes education and the exchange of ideas among those with an interest in the valuation of machinery and equipment. The society, 5000 members strong and headquartered in Washington, D.C., is self-supporting, unaffiliated, and the only organization to represent the entire disciplinary spectrum of appraisal specialists. Its Machinery and Equipment Committee, which collaborated on the writing of this volume, is comprised of fully tested and certified appraisers of machinery and equipment, the leaders in their field.